The Portland Cement Association's Guide to Concrete Homebuilding Systems

The Portland Cement Association's Guide to Concrete Homebuilding Systems

Pieter A. VanderWerf
W. Keith Munsell

McGraw-Hill, Inc.

New York San Francisco Washington, D.C. Auckland Bogotá
Caracas Lisbon London Madrid Mexico City Milan
Montreal New Delhi San Juan Singapore
Sydney Tokyo Toronto

©1995 by **Portland Cement Association.**
Published by McGraw-Hill, Inc.

hc   1 2 3 4 5 6 7 8 9 0 DOH/DOH 9 9 8 7 6 5 4

Library of Congress Cataloging-in-Publication Data
VanderWerf, Pieter A.
    The Portland Cement Association's guide to concrete homebuilding
systems / Pieter A. VanderWerf & W. Keith Munsell.
        p.   cm.
    Includes index.
    ISBN 0-07-067020-X
    1. Concrete houses—Design and construction.   2. Concrete blocks.
3. Gunite.   I. Munsell, W. Keith.   II. Portland Cement Association.
III. Title.
TH4818.C6V36   1994
693'.5—dc20   94-22245
                                                        CIP

Acquisitions editor: April D. Nolan
Editorial team: Robert E. Ostrander, Executive Editor
                Sally Anne Glover, Book Editor
Production team: Katherine G. Brown, Director
                Rhonda E. Baker, Coding
                Donna K. Harlacher, Coding
                Toya B. Warner, Computer Artist
                Rose McFarland, Desktop Operator
                Nancy K. Mickley, Proofreading
                Jodi L. Tyler, Indexer   0070670020X
Designer: Jaclyn J. Boone   GEN3

Contents

Acknowledgments

This project owes its existence to Dan Mistick of the Portland Cement Association. He helped conceive it, he got resources for it when there appeared to be none available, and he gave it a clear focus that he defended against all attempts to dilute it. Several others at PCA were also instrumental in guiding and supporting the effort, notably Bruce McIntosh and George Barney.

The National Association of Home Builders also gave us massive support. The flagbearer for NAHB was Eric Lund of the Research Center, who painstakingly proofed the drafts and made important corrections and valuable suggestions for improvement. We only wish that we could have followed every one of them. Dick Morris of the Technology and Codes Division came up with the original idea of getting our information by interviewing builders, and he read and commented on large sections. Dan Johnson of Technology and Codes gave us useful directions, as well as lots of specific information on code issues.

Scott Ramminger of the National Concrete Masonry Association was a tremendous help on matters related to block. Bob VanLaningham and Bob Thomas of NCMA gave us important technical information.

Paul Brosnahan of the American Institute of Architects read drafts and made useful suggestions from a designer's perspective. David Stevenson, who must be the world's most knowledgeable person on shotcrete construction systems, virtually outlined that section of the book for us.

Elena Mansour, perhaps the finest administrative assistant we've worked with, organized two highly unorganized authors to the point that the writing and production of the book went reasonably smoothly. Rebekah Shub stepped into the breach when we hit bumps and helped make sure the work got done according to plan, often sacrificing her personal time in the process. Mary Jennings and Karyn Wergland put up with hundreds of requests to type, copy, send, fax, call, and answer, and always managed to pull it off with a smile.

April Nolan and Sally Glover of McGraw-Hill were fantastic editors. They consistently received drafts late, yet turned them into polished copy on schedule without complaining.

Shari VanderWerf, a professional writer who also happens to be the wife of one of us, helped to set the writing style. And even when our wives, Shari and Martha, didn't work on the book, they did a marvelous job on the scrambled brains and bruised egos of the harried authors.

Finally, just as builders have to thank the crews who did the real physical work on their projects, the authors wish to thank the hundreds of builders, tradespeople, architects, engineers, building officials, code officials, realtors, home buyers, product manufacturers, laboratory staff, and distributors and retailers who gave their personal time to share their knowledge of concrete home-building systems with us for no personal gain except the satisfaction of helping others. The ones whose names we happened to remember or write down follow. Our apologies to the rest.

Bill Abatte	Hebel Southeast
Teagan Abear	Gilbert Block
Harry Ackner	Ackurate Concrete Company
Arnold Albert	Reddi-Form, Inc.
Lars Anderson	UC Industries
Ron Anders	
Bob Bardell	S&A Custom Built Homes Inc.
Kenneth Baxter	K&M Industries
Ron Benaringsfield	Featherlite Building Products
David Binder	American ConForm Industries
David Bizzel	American Polysteel Forms
Steve Black	Truss Panel Systems
Ken Blake	ThermaLock Products, Inc.
Jeff Bleaman	Bleaman and Associates
Lazlo Bock	American ConForm Industries
Jim Boggus	Corotherm Building Systems, Inc.
John Bongiovanni	Keystone Systems, Inc.
Lance Borrenborg	American Polysteel Forms
Tom Boudreau	City of Newton
Blake Bowthorpe	Bowthorpe Builders
Gunther Bauer	Bauer and Gosch, Inc.
Andy Bradley	
Paul Brosnahan	American Institute of Architects
Markeeta Bundschuh	Pebble Creek Golf Resort
Don Burkett	Burkett Stucco, Inc.
Dennis Calvert	Robson Communities
Dan Campbell	Truestone Block Company
Robert Carmody	Betco Block and Products, Inc.
Steve Cassentini	
Rusty Caudle	New England Air
Mike Cheatham	Mike Cheatham Plastering
Paul Charbonnay	PPC Plumbing
Bill Cherry	Ballantine & Cherry, Inc.
Ron Christanson	
Steve Cobb	EnerG Corp.

Chris Coleman	Morrison Homes
Vic Coleman	Vic Coleman & Sons
Jim Cullinan	Graham Gund Architects
Kevin Cunningham	Dryvit Systems, Inc.
Stephen Cupp	Royall Wall Systems, Inc.
William David	W.D. Contracting Co.
John Decker	Penn Weber Properties
Ward D'Elia	Samyn-D'Elia Architects
Ronald Demerjian	Insurance Institute for Property Loss Reduction
JD Dennis	
Ursula DeVere	Superior Walls of America, Ltd.
Edgar Ducharme	Chicopee Mason Supplies
Dean Ducharme	Ducharme Block
Pat Dunn	UC Industries
Robert Dunning	Dunning Companies
Wayne Fenton	Lite Form
Carlos Ferre	Melton-Ferre Architects
Earl Findley	Advanced Construction Technology
Dennis Finnegan	Watkins Concrete Block
Roy Franklin	
Scott Freeman	Conproco Coatings
Edward Freyermuth	Superlite Block
Bill Furness	Furness Construction
Rick Gale	Rick Gale Construction
Bob Gates	Gates Manufacturing
Drake Gastineau	I.C.E. Block
Ian Geisler	American Conform Industries
Randy Gleave	Rocky Mountain Masonry Systems
Doug Gordon	American Institute of Architects
Leon Graber	Thermo-Tech Insulation Contractors
John Greenert	J.W.G. Management, Ltd.
Ben Griffin	
Allan Griffiths	Sparfil Blok Florida Inc.
John Gross	Rocky Mountain Masonry Systems
Henry Gruber	Hänks Häuser
Jim Gulde	Florida Concrete and Products Association
Jim Hanamaikai	Rocky Mountain Masonry Systems
Rodney Harris	Dallas Harris Real Estate Construction
Peter Harrison	Parex Inc.
Scott Hawkins	
John Hayes	
Carl Henrickson	Keeva International, Inc.
Geir Hjorth	Perma Form
Royce Hobbs	Insteel Construction Systems, Inc.

Charles Hoff	Mansfield Township
Bill Hooderman	Corotherm Building Systems, Inc.
Eddie Horan	
Jerry Horsley	Jerry Horsley Construction, Inc.
John Horton	Ridgeway Building Associates
Tony Ianarelli	Insul Block Corporation
Mike Ibrahim	Hebel Southeast
Stan Johnson	Insulated Masonry Systems, Inc.
Rick Johnson	Insulated Masonry Systems, Inc.
Henry Jorgensen	Larson Cement Stone Company
Doug Kaiser	Sun Valley Masonry, Inc.
Paul Elder Karp	Karp Construction
Russell Kennedy	Kennedy Associates
Francis Kennedy	ThermaLock Products, Inc.
Jim King	King Konstruction
Fritz Kramer	Featherlite Building Products
Pat Kraske	Insta-Foam Products, Inc.
Steve Kroger	West Rivers Masonry
Jerry Kunes	Erie Company
Wayne Lambert	Plantation Construction Company
John Lamere	Energrid, Inc.
Chris Lang	UC Industries
Terry Lapeller	Building Officials & Code Administration International, Inc.
Ken Lewis	Ariel-Triad, Inc.
Dan Lloyd	Covintec-Therml Impac International
Gerald Lundrigan	Lundrigan Construction
Andrew Maizner	Wasatch Solar Construction
John Maloney	BASF Corporation
Bill Manson	
Carl Martin	Superior Walls of America, Ltd.
Bob Martin	C.W. Weaver Inc.
Bill Mason	Therma Manufacturing
Jerry Mazillo	Owen Brook Homes
Paul McArthur	Paul McArthur Builder, Inc.
Kerry McCoole	Parex Incorporated
Julian McDonald	Insteel Construction Systems, Inc.
Clint McGrath	Insulated Masonry Systems, Inc.
Owen McGuinnes	Stylex Homes
Michael McIvor	GREENBLOCK Worldwide Corp.
Jim Meadows	Fox Run Construction Co.
Serge Meilleor	Distribution of Polycrete of Montreal
Henk Mensen	AAB Building Systems
Lineke Menson	AAB Building Systems

Mike Mercer	
Mike Merrigan	International Conference Office of Building Officials
Glen Micale	Micale Environmental Building Systems, Inc.
David Miller	
Denny Miller	Miller Construction
Roger Miller	Thermal Block Systems
Scott Miller	Mike Lilly Construction
Frank Mooney	Lundey Sales Corp.
Mike Morcor	
Bob Morgan	American Society for Testing and Materials
Dick Morgan	Town of Stoughton
John Mumau	Owens-Corning Fiberglass Corporation
Brian Nagle	Robson Communities
Dave Nelson	Pneutek, Inc.
Gordon Nelson	LifeTime Homes
Judy Niemeyer	Tierra Homes
Steve Owenby	Castles of Georgia
Jorge Pardo	Synthesis International Inc.
Carol Parmally	Featherlite Building Products
Bob Patek	The Patek Companies
Sherwood Patek	The Patek Companies
Michael Philippes	
Paul Piergallini	PAG Industries
Eric Plane	AAB Building Systems
Tom Potts	Freedom Enterprises
Jeff Preble	Jeff Preble Concrete
Don Pruss	GREENBLOCK Worldwide Corp.
J.N.C. Rassias	Intralock Corporation
Steve Rhoades	Hebel Southeast
Rick Roach	Barnes and Cone
Lou Robbins	Hilti Inc.
Craig Robinson	
Pierre Andre Rodrigue	Distribution Polycrete of Montreal
David Rossetti	Raven Homes
Dan Rourke	Applied Wall Systems
Charles Rucci	Rucci Construction
Don Sample	SBS Homes
Bob Satter	The Satter Companies
Dan Savant	Angelus Block Company
Don Schmidt	D.L. Schmidt Construction
Allan Schofield	Belknap Construction
Kristin Shultz	Lite-Form, Inc.
Fred Schwab	Morrison Homes
Toby Smith	American ConForm

David Sholes	Rocky Mountain Masonry Sytems
Barry Siadat	W.R. Grace & Company
Gary Sillason	
Don Singer	
Bill Singer	Singer Masonry
Scott Smith	Dow Chemical Company
John Sneep	Majesty Builders
Bill Snow	Palmetto Gunite Company
Dr. Marco Souza	Corev Inc.
Bob Spangler	Conference of American Building Officials
Loren Spies	Loren Spies Construction and Development, Inc.
Tim Spilker	Spilker Masonry
John Stabile	John Stabile and Son
Chuck Stillman	Featherlite Building Products
Bruce Swoyer	Union Carbide Corp.
Paul Tagliero	Cone Constructors, Inc.
Randy Taylor	Morrison Homes
Tom Tedesco	Tacto Electrical Services
Scott Tonnilier	Tonnilier Custom Homes
Kim Ulvested	Thermomass
B.J. Ungvarsky	Gosnell Development Company
Keith Utter	Town of Newton
Joe Van Holland	
John Veciunca	JV Construction, Inc.
Rick Vognild	Southern Building Code Congress International
Randy Ward	American Society of Heating, Refrigeration, and Air Conditioning Engineers
Doug Wattenberger	Calcrete Company
Lloyd West	West Materials, Inc.
John Whittaker	Homes by Whittaker
Bob Wiggins	Carolina Construction Company
Jay Williamson	3-10 Foam Form
Pete Wismann	Calstone Company
Becky Wolf	Hilti, Inc.
Paul Yager	Yager Associates
Frank Young	Sunland Homes
Jerry Yurkoski	Souza, True and Partners, Inc.

Introduction

This book is for building professionals who are interested in learning about (and maybe trying) the *new* concrete homebuilding systems (CHSs). Builder interest in these systems intensified two years ago when lumber prices jumped and builders wanted a cost-effective replacement for structural lumber. But the builders who tried CHSs found that there were reasons to use them other than simply to cut expenses when lumber prices rose. In some situations, CHSs are less expensive than wood frame, even when lumber prices are not particularly high. CHSs are generally more durable, they survive wind and fire better, and they measure up in areas where concrete was traditionally considered weak (market appeal, attractiveness, water resistance, insulation, and ease of construction). Some CHSs are more architecturally striking, more comfortable for the occupants, more energy efficient, and easier to erect than frame. The new CHSs are one of the best-kept secrets in the construction industry, a far cry from the gray block and poured concrete of the past (Fig. I-1).

I-1a *Shotcreting an insulated steel mesh panel.* Insteel Construction Systems, Inc.

xiii

I-1b *Pouring concrete into walls of hollow foam blocks.* GREENBLOCK Worldwide Corp.

I-1c *Building with pre-insulated concrete block.* ThermaLock Products, Inc.

First and foremost, this book is for the general contractor who's looking for new products that will improve business. The book's designed to help builders decide whether the concrete homebuilding systems make good business sense, explain how to choose the CHS that fits the builder best, and show how to get started. The book lists the available systems, describes their design, discusses how well (or how badly) they've worked in the field, and explains where to get detailed technical information. It also covers how the systems have worked out in the market: where they're selling, who's buying them, and the promotion and sales techniques that have moved them the fastest.

The book might also be useful to architects, masonry contractors, and concrete contractors. Residential architects need much the same information that builders do to evaluate and use new systems. The trades need to know enough to bid on jobs specifying CHSs. From the facts in the book and the directory of manufacturers at the end, all of these people should be able to get the information they need.

The key information in this book comes from the field. During 1993 and 1994, we did research for the Portland Cement Association and the National Concrete Masonry Association on the use of concrete in homebuilding. Like this book, the research focused on recently developed systems that use concrete to form the above-grade exterior walls of detached single-family housing. We interviewed more than 200 builders who had used these new CHSs. We also interviewed dozens of realtors who sold them, tradespeople who installed them, and buyers who lived in them. And we visited more than 100 job sites. This gave us the practical data and tips a builder needs in order to make practical decisions. We also got plenty of technical information from the system manufacturers' literature, interviews with the manufacturers, code officials, and testing laboratories.

What we learned in the builder interviews is the most valuable part of this book. It became clear that you can't evaluate a building system solely from the technical information. There are too many important construction details that pop up in the field. And the market's reaction is impossible to predict until you try to sell a house. So we got the final word on costs, performance, advantages, problems, and marketability from the people who have firsthand experience: builders, buyers, tradespeople, and realtors.

Outside experts from four trade associations read drafts of this book to help us make it as accurate as possible. Being accurate isn't easy because many of the items we cover are partly a matter of judgment (like a system's market appeal or the ease of making design changes). Other items like costs vary with a whole host of conditions. The informed eyes of our outside reviewers gave us an independent check on reliability. Naturally, all final decisions were those of the authors, and we take full responsibility for the content of the book.

Getting the most out of this book

If you want to jump right to the descriptions of some of the concrete home-building systems in chapters 5 through 9, go ahead. For most builders, they're just plain fun to learn about.

But when you get serious about considering using a CHS, you'll learn a lot about how to choose the one that's best for you if you're more systematic. Start by reading chapters 1 through 3 straight through. Chapter 1 lays out the trends that have motivated thousands of builders to use the CHSs. It should help you think through whether the CHSs make sense for you and what you want from them before you launch into a full evaluation of all the systems out there.

Chapters 2 and 3 give you enough information about each CHS so you can turn directly to the ones that are most likely to deliver what you're after. Chapter 2 gives a quick description of how each CHS works and what its advantages are. This chapter is also helpful because it shows just how radically different the new CHSs can be from traditional concrete. If you dive into the rest of the book with visions of gray block and plywood form work in your head, you could be hopelessly lost. Chapter 3 tells which systems have been selling to which types of buyers. It lists the different types of customers and the systems that have sold best to each one. After reading chapter 3, you should get ideas about which systems are best for your area and market segment.

At this point you might want to skip chapter 4 and read the detailed descriptions of the systems that interest you in chapters 5–9. Chapter 4 explains the 20-point checklist we use to describe the CHSs and gives technical background so that you'll thoroughly understand all the facts and numbers. But you can pick up how the checklist is organized by reading about a few systems. And a lot of builders know at least some of the technical background. So you can start by reading about the systems you're interested in, then referring to chapter 4 when you're unclear on something.

We recommend three other steps after you read about your favorite systems. First, when you think you've located one that's right for you, make sure you've read about the other similar ones, too. At the very least, read about the other systems in the same chapter as your top choice. This will make you rethink your decision, and since it's a big one, that's a good precaution. And it will give you ideas. Often we describe construction and marketing tricks that have worked successfully with other systems that you could apply to your own as well.

Second, go back and make sure you've read all the parts of chapter 4 at one time or another. Technical background isn't fun to read, but it's important to know. Picking up the parts of chapter 4 that you haven't read yet will make you think precisely about all aspects of the system that interests you.

Third, read chapter 10. It describes the key marketing tricks that the most successful builders used to make sure their CHS houses were a big hit. These ideas might help you do the same. Good luck, and may you, the people you work with, your community, and your customers all benefit from your efforts to build a better house.

1

Why try concrete?

We interviewed a builder in Indiana who put up a house of concrete block in 1993 after doing nothing but frame housing for 20 years. Why? When he bought the lot in 1990, he sent for estimates from his subcontractors. But he decided not to build because of the weak market. When he got new estimates in 1993, the cost of the lumber package had increased $10,000. Dumbfounded, he looked into alternatives. He found: an attractive block that looked like rough brick, an effective waterproofing agent and a high R-value insulation for block. The builder decided to take the risk of building the spec (not presold) house out of these new products. He was not disappointed. His construction costs were $2000 less than the estimates he received for a frame version of the house, and it sold before the roof was on. This example illustrates the two recent trends that have made concrete an increasingly attractive structural material for homes: volatile wood prices and innovation in concrete construction.

Most home builders are painfully aware of the problems wood prices have caused. When they nearly doubled between mid-1992 and mid-1993, builders with fixed contracts lost much of their profits. Those with cost-plus contracts or bids out had some unpleasant explaining to do when they went back to their clients with a revised estimate. Prices have since fallen back somewhat. But even when they are low, the possibility that they will rise again makes lumber risky.

Long-term, there's reason to believe wood prices will continue to rise and fluctuate. If the price run-up of 1992 was simply the result of moves to protect the spotted owl, and federal restrictions on logging caused the increase in late 1993, they might well have been one-time blips. But some forestry experts say that the true underlying problem is that we've been cutting down our high-quality forests faster than they've been growing. If this is true, it can't go on forever. The rate of lumber production will have to slow, and its price will rise. And along the way, we're in for some more big swings up and down as events—sometimes events in a distant corner of the country—constrict supply.

In contrast, the average price of concrete products has not risen by more than four percent in any year since the rapid inflation of the 1970s. Concrete is made of some of the most abundant materials on earth, with estimated supplies measured in centuries.

But in the past, problems with lumber were not enough to make many builders switch to concrete. When builders investigated the concrete systems that were available, there seemed to be no way to get a comparable house at a comparable price, even if the price of lumber was higher than usual. Concrete had all sorts of other costs—finding or training labor with new skills, rigging up some ad hoc ways to waterproof, insulate, and attach a siding that people would like, convincing the building department that the design was acceptable, and paying increased charges from the plumber and electrician. Once the builder factored these in, any saving from less-expensive materials was gone. It made more sense just to ride out the trouble and continue using wood.

But in the last ten years, dozens of innovations in low-rise concrete construction have eliminated or reduced the old problems. These innovations have been packaged into systems that are now competing successfully with wood and often have additional selling points that wood can't match.

This burst of innovation is likely to continue for a few reasons. Manufacturers of the new systems are getting feedback from the field that's giving them ideas for improvements that will make their systems do their jobs better and cut their costs. Several other innovations that we've seen from independent inventors and from Europe haven't yet even hit the market in the United States. Over time, more of them will make it into the available systems.

These two trends—high lumber prices and improvements in concrete systems—have the potential to make CHSs the best option in more and more projects over time. The materials-cost advantage will grow, the problems and costs of using concrete will decline, and, as concrete gets more exposure, its other benefits will get more notice. The list of potential benefits from using concrete is long. We've listed the ones that builders pointed out to us in Table 1-1. Some are traditional; they've been an advantage of concrete forever just because of the properties of the material. And some are emerging; they've been made possible in certain CHSs by the new innovations. But all could give concrete a boost as buyers begin to see and understand the gains.

It's unlikely that frame will ever disappear completely. There will always be some projects for which its unique combination of characteristics does the best job. In remodeling projects, wood might be preferred because of the advantages of using the same materials as used in the original construction. Wood might be the best choice for temporary structures for years to come because it's easier to demolish. But if these trends continue, the number of projects where wood fills the bill will shrink. Over time the predominant material for home construction might switch from wood to concrete, just as it did in European homes during the nineteenth century.

Obviously, these long-term trends do not mean that every builder should switch to concrete construction today. Many builders are in areas where materials availability and buyer preferences will favor frame for years to come. And some are near the end of their careers. For them, learning a new way to build

Table 1-1. Benefits of Concrete Structures for Homes

Traditional	Emerging
Price stability	Low cost
Supply stability	Attractive finishes
Durability	High R-value
Low maintenance	Design flexibility
Fire resistance	Water resistance
Wind resistance	Earthquake resistance
Reduced insurance premiums	Speed of construction
Thermal mass and energy efficiency	Ease of construction
Termite and insect resistance	All-weather construction
Rodent resistance	Low skill requirements
Rot resistance	
Local supply	
Sound resistance	
Nontoxicity	

isn't worth the effort, even if it does mean more jobs or profit for the last couple of years. But if you're a contractor who doesn't fit into these categories, there are good reasons for you to consider concrete now.

The most obvious reason is that you might be able to get an edge in your current market immediately. You might be able to build a house that's attractive to your customers at a lower price than the competition. So you charge less to get more sales, or you keep your price steady and pocket a little extra. Or you might be able to build a house with novel architectural, durability, or energy features that are tough to get with wood, yet appeal to a certain group of buyers in your area. The house should move fast or command a premium.

Another reason to try concrete now is to build sales volume. Imagine that, in addition to frame, you offered concrete homes with novel, useful properties. This specialty could add sales that your company otherwise wouldn't have gotten. And since there would probably be little local competition, the new business could boost your profit.

One more reason to use a new concrete system might be to smooth sales volume fluctuations that result from spikes in lumber prices. When lumber prices jump, customers have balked at buying. But if you've learned a cost-effective concrete system, you can offer a more attractive home price. Concrete's stable cost lets you bid without the fear that profits will be eaten away by rising prices. It could help you win bids and keep building while others are idle.

Finally, you might just want to build a house that gets referrals for your regular business—something striking for a Parade of Homes or that attracts a story in the local paper. A system that hasn't yet been used in your area and has remarkable new properties might get you recognition as the builder to call.

We didn't make up this list of reasons for using a concrete system. These were the reasons that the builders we interviewed gave us when we asked why they started using concrete.

We won't lie to you. Learning a new homebuilding system, like learning anything new, is a lot of work. But if you're like us and you're like most of the builders we interviewed, it's so interesting that it's fun. And if it boosts your business, it can be well worth the effort.

2

What's new

When we interviewed builders who had used one of the new concrete home-building systems, we sometimes took a few minutes at the end to describe the other systems now available. In almost every case, the builders sat there dumbfounded. Even though these people had each used one of the systems, they had no idea that so many different methods for making houses out of concrete existed. They peppered us with questions until they saw that, yes, these things must really exist and could really work.

The new CHSs are about as close to the plain gray block and concrete forms of 20 years ago as computers are to slide rules. The old concrete houses (almost all of them made of block) gradually fell out of favor after World War II. Compared with frame, they let a lot of moisture inside, were hard to insulate, and their appearance (square, with sharp mortar joints dividing the walls into 8-x-16 rectangles) came to be considered a "cheap" look. Without adding some very expensive extra components to the wall, block simply could not provide a structure that was as dry, well-insulated, and attractive as frame.

But in the last 20 years, hundreds of innovations in concrete construction have found their way into systems suitable for homes. Together these innovations have made systems as dry, energy-efficient, and attractive as frame, all at a similar cost. And all of them are superior to frame in many ways.

Table 2-1 contains quick facts on all of the new CHSs in the book. To keep ourselves focused, we only included the systems that have been used by independent builders to construct a substantial number of detached single-family homes within the last two years (although most have been used in multifamily construction as well). Almost all of them are clever combinations of concrete, steel, plastic foam, and special agents mixed into the concrete. The concrete bears loads, the steel holds against lifting and side forces, the foam provides insulation, and the agents give the concrete color, texture, water resistance, or insulating properties. Yet the different systems use these basic ingredients in such different ways that we have to split them up to describe them.

We divided them into five categories according to the way they apply their concrete: mortared block systems, mortarless block systems, poured-in-place systems, shotcrete systems, and panelized systems.

Table 2-1. The New Concrete Homebuilding Systems

System	Key Components	Distinctive Properties
Mortared Block Systems		
Interior-insulated	Insulation mounted on interior face Waterproofing additives or coatings Decorative block or stucco exterior	Moderate cost Novel aesthetics
Cavity-insulated	Insulation in the middle of the block Waterproofing additives or coatings Decorative or plain block surface on both sides	Low cost or novel aesthetics
Exterior-insulated	Insulation board mounted on outside face Stucco-like exterior	Energy efficiency
Hebel Wall System	Solid blocks Special "aerated" concrete Special high-adhesion mortar	Insulation and overall material quality Cuts and fastens like wood
Integra Wall System	Groutless reinforcing Foam insulation in the middle of the block	Moderate cost High R-value High strength
Sun Block	Special foam and concrete mixture	Built-in insulation Cuts and fastens like wood
Mortarless Block Systems		
IMSI	Stacking with shims Insulation in the middle of the block Surface bonding cement on both sides	Moderate cost Good energy efficiency Low skill requirements
Intralock	Stacking without shims Bonding with a continuous core of grout Spare row of cavities inside and out	Moderate cost Unusual strength Low skill requirements
Sparfil Wall System II	Stacking with few shims Special foam and concrete mixture Special surface bonding system	High energy efficiency Cuts and fastens like wood Low skill requirements
Poured-in-place Systems		
Stay-in-place panels	Panels of foam sheet and plastic ties	High R-value

System	Key Components	Distinctive Properties
	Erected in large sections	Light work
	Cavities filled with concrete	Cuts easily
Stay-in-place grid blocks	Foam blocks	High R-value
	Teeth for aligning	Easy stacking
	Cavities filled with concrete	Cuts easily
Shotcrete Systems		
3-D	Panels of heavy wire mesh	Light work
	Foam sheet core	Cuts and shapes easily
	Concrete applied to mesh on each side	Unusual strength
Panelized Systems		
Royall Wall System	Prefabricated walls	Rapid site assembly
	Inside face of insulation and steel studs	Good design flexibility
	Bolt-together assembly	Low cost
Superior Walls	Prefabricated walls	Rapid site assembly
	Built-in insulation and concrete studs	Simplicity
	Bolt-together assembly	Low cost

We put the mortared block systems first because they're closest to current practice and easiest to understand. Some of them are made simply by adding some new products to the traditional ones. These systems have no separate brand name of their own because no one manufacturer "owns" them. Systems made of off-the-shelf components like this are called nonproprietary systems. Then come the proprietary block systems (ones that are designed, owned, and sold by a single manufacturer) and the systems in the other categories.

Warning: The descriptions in this chapter are incomplete. They're meant only to tell you enough about each system so you can see what's new about it and start to home in on the ones of most interest to you. Chapters 5 through 9 contain the details on the systems.

Mortared block systems

The problems of moisture, low insulation, and bland appearance plagued the old block houses. There are two traditional ways to solve these problems, but both are expensive. The first is the *double-wythe wall*, which is really two separate walls of masonry with a gap between them. Usually the outer wall is brick, and the inner one is block. Moisture that gets past the brick trickles down the gap and out through weep holes. Rigid insulation can go in the middle. And the brick exterior is more than acceptable in most regions of the country. So the system does its job well. But the construction of two separate walls adds so much cost that double-wythe walls have pretty much been confined to the commercial and very high-end residential markets.

The other solution is to add studs to the inside of the block. This separates the interior sheetrock from moisture and leaves room for insulation. But it really doesn't solve the moisture problem. It only hides it from view, and there's nothing in this design to improve the exterior appearance. And, again, this system is costly. It's almost like building two walls again, a stud wall inside a block one.

The new mortared block systems solve the basic problems of moisture, insulation, and appearance in different ways, and they throw in some added features as well. We found three nonproprietary systems and three proprietary ones.

The first of the nonproprietary systems we call *interior-insulated mortared block*. It makes use of water-repellent coatings or agents in the blocks and mortar to keep the wall dry. Inexpensive new studding or metal clip systems that mount insulation directly to the inside surface can increase R-value to levels high enough for almost any climate. New stucco products and striking new types of colored and textured block make a range of striking appearances available. Figure 2-1 shows an interior-insulated block house with a textured block exterior. And some builders told us that the cost of the system in their projects was dead even with frame, or even a little lower.

2-1 *A new split-face block house near Indianapolis.* W.R.Grace & Co.

Cavity-insulated block uses similar agents, stuccoes, and blocks for water resistance and appearance. The novel feature of this system is the use of insulations that go into the cavities of the block: sprayed-in foams, poured-in fills, and rigid foam inserts. Figures 2-2 and 2-3 show a couple of these. They make it possible to leave the block exposed on both outside and inside, which can either save money (and appeal to the low-end market) or produce some remarkable appearances (and appeal to the high end).

2-2 *Installation of sprayed-in foam insulation.* Polymaster, Inc.

2-3 *A specialty pre-insulated concrete block.* Insul Block Corporation

Exterior-insulated block uses the new foam-and-stucco siding products outside a block wall to form a water barrier, insulate the block, and provide a sharp appearance. Figure 2-4 shows one of these products being applied. Although the special components add a few thousand dollars to the cost of a typical house, the resulting energy performance is exceptional. A nearly unbroken layer of insulation surrounds a wall of high thermal mass (thermal mass is the ability to absorb and hold large quantities of heat or cold). This is ideal for storing heat in passive solar homes or keeping houses cool with only nighttime air conditioning. Energy-conscious individuals and utilities are the most frequent customers.

2-4 *Attachment of exterior insulation system to a block wall.* Keystone Systems, Inc.

The first proprietary mortared block system we list is the Hebel Wall System. Its solid blocks are made of a revolutionary new form of concrete that's full of millions of tiny air bubbles. The material is light, cuts easily, insulates, and resists water. Figure 2-4a shows a worker stacking Hebel blocks. The manufacturer also supplies a special adhesive mortar and lightweight stucco to bind and seal the blocks. For a little more money, the finished wall performs better than ordinary frame on almost every dimension important to quality-conscious buyers.

The Integra Wall System employs special blocks with a minimum of concrete in the center and special tensioned reinforcing that clamps the wall down without grout. Instead of concrete and grout, the center of the wall is filled with a foam that insulates and seals it, as shown in Figure 2-4b. For a cost that's about even with frame, the result is a high-R, high-strength structure with the other advantages of concrete.

2-4a *Laying Hebel block.* Hebel Southeast

2-4b *Injection of foam insulation into an Integra wall.*
Superlite Block

Sun Block is made of a patented mixture of concrete and foam beads. Figure 2-5 shows one of the blocks. Because Sun Blocks are shaped and stacked like conventional blocks, learning to use them is easy. But the material also makes them light, about as easy to cut and fasten to as wood, and good insulators. For builders accustomed to frame, they're "block made easy." For buyers, Sun Blocks provide a home with additional insulation and the quality of concrete for a little extra money.

2-5 *A Sun Block.* Sparfil Blok Florida

Mortarless block systems

The mortarless systems use concrete blocks that stack directly on top of one another without mortar. Once stacked, they're bonded with materials troweled onto or poured inside of the wall. This sharply reduces the time and skill required for erecting the structure, while sealing the entire envelope. But each of the three mortarless block systems also has additional important features.

IMSI (named after Insulated Masonry Systems, Inc., the manufacturer) is bonded by troweling onto both faces a fiber-reinforced surface bonding material that gives strength and resistance to water and air, as well as a finished surface. IMSI blocks, shown in Fig. 2-6, have several cavities that provide space for rigid foam insulation, steel reinforcing, and utility lines. For the same cost as frame in many situations, the balanced design of the IMSI system provides a wall

2-6 *The parts of an IMSI block wall.* Insulated Masonry Systems, Inc.

with water resistance, an attractive finish, and an R-value that's high enough for northern climates.

Intralock blocks are so precisely sized that they stack without shims. They also have a narrow core that's filled with grout after stacking to form a continuous internal slab of reinforced concrete, shown in Fig. 2-7. There are also extra cavities that can be used for insulation and utility lines. The exterior is usually finished with one of the various stucco products. For about the same cost as frame, the final structure is exceptionally tight and strong.

Sparfil Wall System II uses blocks made of the same foam-and-cement mixture as Sun Block. Sparfil blocks are lightweight and cut like wood. But they're also sized to stack with little shimming, and they have a system of cavities and rigid foam inserts that give the final wall an R-value of more than 20. That and the thermal mass of the concrete produce unusually high energy efficiency at a little more than the cost of frame. Figure 2-8 shows the block and its inserts. Both the exterior and interior surface are covered with a troweled material and a fiberglass mesh to bond and seal the wall.

2-7 *Intralock wall cut away to reveal sections of its interior slab of reinforced grout.* Intralock Corporation

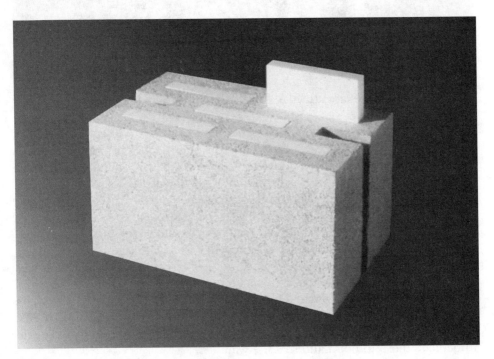

2-8 *Sparfil block with inserts.* Sparfil International, Inc.

Poured-in-place systems

Poured-in-place reinforced concrete has long been used to form exterior walls in commercial construction, but rarely single-family homes. As with block, the poured walls could be difficult to insulate, prone to transmitting water, and difficult to side attractively.

Traditional poured-in-place concrete is also not very economical in most single-family projects. Setting the forms is expensive enough that you have to spread the cost out over a large structure or many similar units to make it pay. Even slight variations from one house to the next (such as moving windows) traditionally requires some expensive resetting of forms.

But all of this is changing with the so-called stay-in-place systems. Simply put, they're concrete forms made of plastic foam. The crews stack them in the shape of the walls and fill them with rebar and then concrete. Afterward, the foam forms stay (which is where the name comes from) to act as insulation and an air and vapor barrier. Interesting exterior trim and moldings are easily made by carving or adding onto the exterior foam and covering with various troweled sidings. With some brands of these forms, screwed and nailed sidings are also easy to put on.

Yet the stay-in-place systems are also close to the price of frame, even in small volume. The forms cost more than lumber, but they provide insulation and sheathing in addition to forming the structure. Cutting and setting them is much like carpentry, except that the work is lighter. For all of these reasons, the cost of labor is reduced and partially offsets the materials cost. There are now dozens of brands of stay-in-place forms. We've divided them into two categories: the panel systems and the grid systems.

The stay-in-place panel systems use off-the-shelf foam sheet. Two sheets are held a constant distance apart by plastic ties, as shown in Fig. 2-9. This produces a concrete wall of constant thickness, just like conventional forms, so it looks familiar to designers, trades, and code officials. What's new, of course, is that the final wall is faced with 2 inches of foam on either side that insulates, backs various sidings, and can be cut away to form secure channels for plumbing and electrical lines. Many builders stucco directly over the foam, but it's also possible to screw into the ends of the plastic ties to attach other sidings. The insulation is about R-20, the wall is relatively airtight and watertight, and it has the strength of reinforced concrete.

Instead of sheet foam, the stay-in-place grid systems use specially molded "blocks" of foam. Most of the brands include teeth or ridges along the edges of the block so they align and lock into place like Lego blocks. Figure 2-10 shows a cutaway view of one grid wall, and Fig. 2-11 shows an actual wall in progress. Inside their cavities, the blocks aren't flat. They have a wavy surface or set of closed cavities for the concrete to flow in, forming a sort of "grid" of concrete. Although the block is more expensive than sheet stock, it goes up quickly, requires little bracing, and the grid shape requires a little less concrete while still retaining strength. The resulting wall often has an R-value of 20 or more.

2-9 *Walls of R-Forms
stay-in-place panels.* R-FORMS, Inc.

2-10 *Cutaway diagram of a
Featherlite grid wall.*
Featherlite Building Products

2-11 *Structure of Reddi-Form stay-in-place grid blocks in progress.* Reddi-Form, Inc.

Shotcrete systems

Concrete shot or "sprayed" under pressure from a pump and nozzle onto a steel framework has been used in the United States for years to build industrial structures such as grain silos. The process has great flexibility. It can produce irregular curves, angles, and openings for relatively little extra labor and with little waste material. The end structure also has tremendous strength.

The one shotcrete system that's currently marketed to home builders in the United States is 3-D. The basic "building block" of the system is the 3-D panel, which consists of a sheet of foam sandwiched between two layers of heavy-gauge steel mesh. Crews erect the panels in the shape of the house's exterior walls, as shown in Fig. 2-12, then spray it on both sides. Finishers can texture the concrete directly or apply various troweled sidings that also provide water resistance. The panels come with different thicknesses of foam that go up to high R-values. The concrete faces form a seamless barrier inside and out. And the final structure is exceptionally strong, all at somewhat greater cost than frame construction.

Panelized systems

Casting entire walls in a plant or on a slab on-site with a pre-built mold is another longtime commercial technology that, through innovation, has recently

2-12 *Installation of 3-D panels.* Insteel Construction Systems, Inc.

become practical for residential construction. Making walls in controlled conditions can provide consistent high quality, cut labor costs, eliminate most weather delays, and cut site time to a fraction of the usual length. But in the past, the labor and materials required to set up a casting plant or molding station to produce a particular design of wall meant that (as with traditional poured-in-place systems) production of panels was only economical when the same basic design could be produced in large volume. And the resulting wall often had the other shortcomings of a lot of traditional concrete systems: water penetration, a plain appearance, and no easy way to add insulation.

But improved molds and casting techniques now make it possible to produce walls in small volume at a competitive cost. Added agents and high-density concretes provide water resistance to the new panels. Rigid foams are cast onto the concrete to provide a layer of insulation. New troweled sidings and traditional nailed or screwed sidings (practical with some of these systems) provide an appearance that's popular with many buyers.

There are two panelized systems currently readily available to home-builders in the United States. The Royall Wall System panels have a continuous concrete face on the outside. On the inside they have a sheet of foam interrupted only by metal studs and plates. Openings and raceways for electrical wiring are built into the panels, which are custom-made for each house. The panels use watertight concrete that's textured on the outside for good stucco adhesion. The foam provides insulation that exceeds even most northern code re-

2-13 *Installing Superior Walls on a gravel footing.* Superior Walls of America, Ltd.

quirements. The panels can be set up in less than a day, all for a cost that, in most cases, is a little lower than frame's.

Superior Walls were originally designed for basements, but some builders have begun using them for the rest of the house above grade as well. They sandwich a layer of foam between a continuous face of watertight concrete (the exterior surface) and a set of concrete "studs" (on the inside). A crew sets them in place with a crane and bolts them together in less than a day to form the house structure, (as shown in Fig. 2-13). It's possible to add troweled sidings or order the panels with optional furring strips to attach other sidings. The concrete studs on the interior have preformed holes for plumbing and attached nailers for sheetrock, trim, and fixtures. The panels provide a simple, high-quality wall that costs almost as little as 2-x-4 construction in some projects.

On the horizon

In addition to the established systems just described, there are several that are so new they haven't been used by independent builders in even a dozen houses yet. This made it nearly impossible to interview experienced builders about them. But because they appear promising, we've included brief descriptions and manufacturers' addresses for getting more information. You'll find them at the back of chapters 5–9.

3

Finding the system that matches your market segment

Asking, "What's the best homebuilding system?" is like asking, "What's the best car?" It depends on the buyer. The one that gives the features the buyer wants at the lowest cost is the one that's best suited to that buyer.

Consider cars. A commuter who wants nothing more than basic transportation could be happy with a Geo Metro. It gets the commuter to work at a lower cost than other models. But if the buyer is a father who also wants room and safety for an entire family, the Metro might not be the best option. He could try to make the Metro work by paying extra for four doors and air bags. To get more room, he could take it to a custom body shop to get it extended. But after he figures in the cost and risk of this extreme amount of modification, he'll probably be better off getting a Taurus. It's not that it's impossible to turn the Metro into a family car. It's just that to do it you drive the total cost way up. It's cheaper and safer to get a model like the Taurus that has the features he wants as standard equipment.

Now suppose the buyer is a mother in a cold climate who wants room and safety, but she also wants good handling in the snow. She might buy a Taurus and pay for optional four-wheel drive. That will help, but with the low ground clearance of a sedan, she'll still have trouble in deep snow. She could go to the next step of having a mechanic jack up the suspension. But if she really wants exceptional handling in high snow, she would probably get it less expensively by buying something like a Jeep Grand Cherokee. True, the base price of the Jeep is higher than the base price of the Taurus. And it's not impossible to meet the buyer's needs with a Taurus. But the Jeep's design fits the buyer's needs with less messy, costly modification. Once you add all the features necessary to make the Taurus ready for deep snow, the Jeep might end up being less expensive and risky.

The same is true of homebuilding systems. The "best" one depends on the features the buyer wants. For someone who wants to own nothing more than basic shelter, a mobile home is hard to beat. It underprices just about everything else available. But another buyer who wants a lot of room and a more durable structure might better switch to wood frame, at least at the lumber prices that existed three years ago. It's not that it's impossible to get these features with a mobile home. You could choose the strongest brand available, piece together a triple-wide, tie it to a foundation, and replace the usual flat roofs with a pitched one to shed rain and snow off the larger combined area. But at traditional wood prices, conventional construction should give the same features less expensively and without the risk that comes from making unusual, untried modifications on the mobile home.

But now suppose the buyer also wants a high degree of fire and wind resistance. Even at old wood prices, the best system might shift to concrete block. You could try to stick with frame. You could add a cement-based siding outside and ¾-inch sheetrock inside for fire resistance. You could attach steel plates at the connections between studs and plates to keep the nails from pulling out in the wind. It's not impossible to get the features the buyer wants with frame. But because block offers fire and wind resistance without much modification, it's likely to be the less expensive system. So, as with cars, the best homebuilding system to use depends on the preferences of the buyer.

With higher lumber prices and improvements in concrete construction, there are now more and more situations in which one of the concrete homebuilding systems provides what the buyer wants at the lowest cost. But which CHS best matches the customer's needs varies. Each system has its own particular features that appeal to different types of people.

Sooner or later you'll have to grapple with the question, "What kind of buyer do *I* want to sell to?" Then you can pinpoint the system that meets the preferences of your buyers most cost-effectively.

The builders we interviewed gave us some insights that should help you mull this over as you read about the new CHSs. The builders described the types of buyers who chose each of the systems. These different types of buyers fall pretty neatly into five market segments. By a market segment we mean a group of people who have similar preferences.

These five market segments are listed in Table 3-1. The table also includes what we learned about which systems sell to which segments. For each CHS, we've marked the segments where the CHS has already sold successfully. We've also marked the other segments where no one has tried to sell the CHS yet (so far as we know), but the CHS seems to fit well.

The five market segments the builders described to us are familiar to most builders. In fact, you probably sell to one or two of these types of customers right now. Let's cover the different segments quickly so you know exactly who's included in each one.

Table 3-1 The CHSs That Have Been Successful in Each Segment of the Home-buying Market

CHS	Value	Quality	Cost	Safety	Energy
Mortared block systems					
Interior-insulated	◉			◉	
Cavity-insulated		◉	◉	0	
Exterior-insulated		0		0	◉
Hebel wall system		◉		0	0
Integra Wall System	◉	◉		0	◉
Sun Block		◉		0	◉
Mortarless block systems					
IMSI	◉	◉	◉	0	0
Intralock	0	◉		◉	
Sparfil		◉	◉	0	◉
Poured-in-place systems					
Stay-in-place panel	◉	◉		0	◉
Stay-in place grid		◉		0	◉
Shotcrete systems					
3-D		◉		◉	◉
Panelized systems					
Royall Wall System	◉	◉	0	◉	0
Superior Walls	◉	◉		0	

◉ Buyers in this segment have selected and bought this system.

0 In the authors' judgement, the features preferred by buyers in this segment are well matched by this system, but they have no examples yet of a builder selling it to them.

The value segment

The value segment is probably 60–70 percent of the total market for new single-family homes. Its buyers are people whose primary concern is getting the most rooms and/or square footage they can afford in a house that looks and works about like the others in the neighborhood. Or, as one builder put it, "They want to stretch their budget into the biggest possible house, so they turn down the fancy extras, but they're too scared of getting junk and too scared about resale to strip it down, either. In the end they usually decide that they want things to look about the same as they are in the neighbors' houses."

These buyers might actually be interested in premium features like solar heating or dramatic architectural details. But they can only spend so much, and when push comes to shove they're not willing to settle for a smaller house in order to pay for these things. Builders said that first-time buyers, people moving out of old construction or poor neighborhoods, and people with little interest in technical matters are almost always value segment buyers.

Builders gave us a list of eight market-related features that they felt were important for any homebuilding system. When we asked them to rank each according to how important it was to value buyers, their answers came out like this:

1 Cost.

2 Exterior and interior finishes.

3 Design flexibility, R-value and energy efficiency, water resistance, maintenance required, and disaster resistance (of about equal importance).

4 Thermal mass.

But they warned us that all of these features rank very close for a value buyer. Reducing total cost is a top concern, but the buyers will go ahead and pay to avoid lowering the quality of the house below standard on almost any of the other features. Value buyers are especially sensitive about the appearance, so the finishes rank high. They care about design and some other premium features, but won't pay as much for them.

The CHSs that have sold to the value segment so far are interior-insulated mortared block, the Integra Wall System, IMSI, the stay-in-place panel systems, the Royall Wall System, and Superior Walls. These can all be built in low-cost, low-frill configurations. Yet they all take popular finishes and high levels of insulation economically. And their quality levels on the other features mentioned by builders can generally be made to match or exceed frame without undue expense. Intralock also appears to have the right features to appeal to the value segment, at least in areas where high R-values are not required.

The quality segment

In contrast to value buyers, there's a sizable group of people who want their home to be different, and they're willing to pay for it. They tend, of course, to be middle to upper income. Almost everyone has his or her own ideas of what's attractive or comfortable or useful. But the wealthy can also pay for the unusual materials and custom work necessary to build their ideas into their homes. And they're less concerned about the house's appeal to others. Being different is part of the attraction, even if it costs something at resale time. These people tend to be older, so they have some money saved up and don't have to worry about pleasing the kids anymore.

The builders said that this type of buyer ranks a wall's features about like this:

1 Exterior and interior finishes, design flexibility, R-value and energy efficiency, thermal mass, water resistance, maintenance required, disaster resistance.

2 Cost.

In other words, the buyers are willing to pay more for premium features of almost every type. It's true that the importance of each feature will vary from person to person. Someone interested in a dramatic architectural appearance will place a high priority on finishes and design flexibility and care less about the other features. Someone interested in energy will look for energy efficiency and thermal mass, and enthusiasts for durability and quality of construction will put a big priority on water resistance, maintenance, and disaster resistance. But, generally, a quality buyer wants some improvement over standard construction on all of these features. It's rare that a buyer who's a big fan of striking architecture and who pays a big premium to get just the right look will settle for a house with ordinary insulation and durability.

Almost every CHS has sold to quality market buyers. A major reason for their appeal in this segment is that many of the features important to quality buyers come automatically with concrete; things like high fire resistance, thermal mass, and low maintenance are inherent in the material, as already listed in Table 1-1.

The mortared and mortarless block systems might seem to be at a disadvantage nonetheless because they're limited in their design flexibility. They generally have a hard time making anything but right angles and square openings. But the Hebel wall system, Sparfil, and Sun Block are highly flexible, sometimes more so than lumber. And most of the other types of block are not so constrained as frame builders tend to assume, as you'll see when you read the sections on block. Some of them also make up for the flexibility limitation with novel exterior and interior finishes that aren't available on frame. As one builder of large custom homes put it, "The thing about block is that it's different. We've been giving people frame for so long now that they're getting tired of the things we can do with it. Block has a lot of looks that people haven't seen before."

The stay-in-place panel and grid systems, 3-D panels, and the Royall Wall System are capable of some amazing curves, angles, and openings at relatively modest additional cost. The advantage the poured-in-place and 3-D systems have is that they are quickly and easily cut and reassembled into unorthodox shapes. The Royall Wall System uses lightweight forming techniques to produce custom shapes quickly.

Energy is the other area where concrete has sometimes come up short of the quality buyer's expectations. But nearly all of the CHSs use new tricks to add insulation less expensively to concrete than has been done in the past. And all use their thermal mass to good advantage, leveling temperatures and reducing total fuel bills below those of comparably insulated frame houses.

The cost segment

Some buyers will sacrifice the standard amenities for the sake of saving money. These are mostly low-income people and people putting up a vacation or second home.

Low-income buyers might be barely able to afford any permanent structure. So they look for the least expensive possible house, and they're willing to sacrifice good looks and conveniences to get it. Builders we met who sell to low-income buyers use the least expensive system for constructing a basic house, sometimes even if the system is a new and different one.

It seemed odd to us at first, but we learned that some people buying vacation or second homes are interested in most of the same features the low-income are. Wealthy and middle-income people sometimes put up getaway houses in rustic areas—in the mountains, in the forest, or on a distant coast. They don't want to spend as much for it as they did for their primary residence because they won't use it as often. And besides, "roughing it" is part of the vacation experience.

When we asked builders to tell us how the key features of a wall ranked in importance for these buyers, we got something like this:

1 Cost.

2 Cost.

3 Cost.

4 R-value and energy efficiency, water resistance, maintenance required, disaster resistance.

5 Exterior and interior finishes, design flexibility, thermal mass.

Cost buyers, we're told, will sacrifice far more than other buyers to save money. Energy efficiency, water resistance, maintenance, and disaster resistance are somewhat important to those building a vacation home. This is because the home is often in a rural area where it's subject to extreme conditions, and the owner might not want to care for it much or spend money on it. Low-income buyers might have similar concerns and might sometimes pay more for these same features, but usually can't.

Cavity-insulated block, IMSI, Sparfil, and the panelized systems all strip down well and have been sold to this market segment. The common feature of these systems is that once they're erected, they can get away with little or no finish work.

In contrast, you usually can't strip away as much of a frame wall and still get a sound structure. Exterior sheathing or bracing has to go on to resist racking forces. Some form of siding must be added to keep moisture out. Some interior sheathing has to cover the cavity.

The safety segment

Some buyers want a house that can stand up to disasters like fires, hurricanes, and earthquakes, and they're willing to pay more for it. Unlike quality buyers, they might have little concern for other premium features. They might accept a house that's a little more plain or higher in energy consumption if it means a stronger house.

The priorities for these buyers tend to be something like:
1 Disaster resistance.
2 Water resistance.
3 Maintenance required.
4 Cost.
5 Exterior and interior finishes, design flexibility, R-value and energy efficiency, thermal mass.

Anything related to survival and durability is at the top.

Builders have successfully pitched the fire and wind resistance of block for years. But almost every other CHS is also sharply more resistant to these disasters than frame is. The underlying reason for this is that concrete itself doesn't burn, and its weight counters the uplift of wind.

The question mark about concrete in the past has been whether it could match frame in an earthquake. But the reinforcing schemes used in the new CHSs generally appear up to the task. The Integra Wall System, IMSI, and 3-D all feature beefed-up reinforcing designed to withstand seismic forces, and they're getting some of their sales because of it. Many of the other CHSs can take extra reinforcing at little extra cost, which should make them relatively earthquake-resistant inexpensively. These include Intralock and all of the poured-in-place and panelized systems.

The energy segment

Some people have become sensitive to the energy efficiency of their homes, and they're willing to pay more for it. And they're often willing to put up with inconveniences and odd-looking architecture if need be. They feel this way because they want to save the money they spend on fuel, they're upset about the United States importing so much fuel, they're environmentally sensitive, or they simply enjoy technology. As one homeowner said to us, "I hate the idea of giving all that money to the utility each month." Another said, "I think this stuff is the greatest. It's really neat."

The priorities of these buyers, as builders reported them to us, were as follows:
1 R-value and energy efficiency.
2 Thermal mass.
3 Cost.
4 Exterior and interior finishes, design flexibility, water resistance, maintenance required, disaster resistance.

The inherent advantage of concrete construction in this market segment is its high thermal mass. This mass evens out heating and cooling peaks and (when it's inside the insulation) stores heat or coolness. Getting that in a frame house requires considerable additional expense. In addition, the past problems of putting high insulation in a concrete wall have pretty well been solved.

The systems that we know to have sold based on their energy features are exterior-insulated block, the Integra Wall System, Sun Block, Sparfil, and most of the stay-in-place poured systems. All of these except Sun Block have more insulation than even most northern codes require. A few exceed R-25 with no special measures. Sun Block is a more modest R-8, but it has shown its appeal in Southern climates. Most of these also have relatively great amounts of concrete (providing thermal mass) inside the insulation. Several other systems, noted in Table 3-1, appear to have a similar set of features. They might also appeal to the energy segment if they were promoted there.

Next steps

At this point you probably have some ideas about the types of buyers you want to sell to and which systems would fit them the best. But there's one major consideration still to take into account. You're not the buyer. You're the seller. To use our car example again, you're not just trying to find a car that you think people will want to buy, you're trying to find a line of cars for which you want to be a dealer.

This is more complicated than just picking a car that has the features people in your area want. Consider, for example, a car dealer in Aspen, Colorado. He might choose to sell Range Rovers because there are a lot of people in the area who want snow handling and high luxury and have the money to pay for it. But because he's the dealer, other features of the brand (such as availability and logistics) figure in. If Range Rover's distribution were weak in the Rocky Mountains, he might have trouble getting cars to sell on time. If the Range Rover's design were unusually complex, his mechanics might have trouble doing their dealer prep and repairs. It wouldn't matter that the Range Rover's features fit his buyers like a glove if he couldn't get enough into the showroom and the buyers ended up dissatisfied because his service was poor. So he might settle for a Jeep dealership instead. The Jeep's features might not fit what his particular buyers want as exactly as the Range Rover's. But if he can provide and service the Jeep more reliably and inexpensively, it might make a more successful business anyway.

As a "dealer" of homes, you have to consider the same types of availability and logistics features when you choose a homebuilding system. It's important to know that a particular system provides most of the features your buyers want at a low cost. But that's not enough. If you can't get the parts or labor, or learning and using the system is difficult, you might be better off picking a system that's not quite as perfect a match to your buyers' preferences but is easier to handle from your end.

The rest of this book covers all of these different types of features for each CHS. Chapters 5–9 contain a section on each system, covering cost, market appeal, availability, and logistics. Chapter 4 gives the technical background on each of these items, which you can refer to as needed. Between them you should be able to weigh all features and make an intelligent choice of one or two systems that could work for you.

4

Background

When we started this book, we asked builders what they'd want to know about a new CHS to decide whether to try it or not. They named a couple of dozen of concerns that fell into four basic questions: How much does it cost? What does the buyer get for the money? How easy is it to get the parts and people to put it together? How do you handle the tasks that are tricky with new products (like getting building department approval) or tasks that were difficult with the old concrete systems (like moving a window)?

So we took all the things they wanted to know about and put them into four main categories. The result was this list of 20 features of a CHS:

- Cost.
  - **1** Labor and materials cost.
  - **2** Learning costs.
- Market appeal.
  - **3** Buyer reaction.
  - **4** Exterior and interior finishes.
  - **5** Design flexibility.
  - **6** R-value and energy efficiency.
  - **7** Thermal mass.
  - **8** Water resistance.
  - **9** Maintenance required.
  - **10** Disaster resistance.
- Availability.
  - **11** Product availability.
  - **12** Labor availability.
- Logistics.
  - **13** Building department approval.
  - **14** Required calendar time.
  - **15** Crew coordination.
  - **16** Making connections.
  - **17** Utilities installation.
  - **18** Change flexibility.

19 Manufacturer support.

- Other.

20 Other considerations.

Chapters 5–9 describe all of the new CHSs, one by one. Each system description has a brief overview, then 20 sections that tell how the system rates on each of these 20 features. Right now it might not be clear what each of these features is, but it will become clear as you read about them. In fact, we've already mentioned the cost and market appeal features in chapter 3 when we discussed what things are most important to each market segment. The feature labeled "other" is a catchall. While studying the CHSs, we stumbled across a few important matters that no one had asked about before we started. So we added this category to have a place to cover them.

This chapter gives background on each of the 20 features. A lot of it is technical background—physical properties of concrete, how to interpret advertised R-values, and so on. The rest explains our use of terms and tells what kinds of information we provide under each feature when we discuss it for the separate CHSs. All of this background is meant to help you understand the descriptions in chapters 5–9.

Feel free to skip this chapter if you want. You might already know a lot of what it covers. And it's not nearly as fun to read as the descriptions of the new systems. But if you find you don't understand something we say about one of the CHSs, the explanation is probably in this chapter. Just flip back and read the background on whatever's confusing you. And it's a good idea before you finish to make sure you've read all of chapter 4 at one time or another. It's easy to think you know everything about a topic when you really don't. R-values are a great example. If you think you built an R-11 wall last week, you're probably wrong. We show you here why you're wrong and about how far off you are. In the process we show you how to interpret the energy efficiency numbers for the CHSs and compare them accurately to frame. This kind of detailed background on the 20 features is useful to have before you make a final decision on which system to use.

Cost

When builders first told us what they wanted to know about costs, there was a lot of confusion because no one was distinguishing between the costs of using a new system the first time and the costs of using it after the builder and his crews became familiar with it. As we all know, the cost of using any new building component tends to come down with practice. So we split the cost feature into:

- Total cost at smooth operation.
- How much extra time and expense went into getting to smooth operation.

Labor and materials' costs

This section of the system descriptions tells how to estimate the cost of building with a new CHS once you and your crews have used it for a while. Let's be honest: Predicting construction costs is fiendishly difficult. The total cost of using a building component—and especially one as complex as a structural system—is the sum of thousands of smaller costs that change over time, that change from region to region, and that change with a lot of specifics of the site, the design, the crews, and the weather. You can never be sure how much detail you have to take into account to get an accurate estimate.

But accurate cost information is also crucial. If your cost estimates for a CHS are too high, you could avoid it and miss out on a great opportunity. Or you could use it but bid jobs too high and lose them. If your estimates are too low you could underbid and lose your shirt. So in our interviews we pushed the builders using CHSs for cost figures. We decided that even approximate numbers are more useful than none at all.

The builders gave us a few different types of numbers. The first was the overall cost of using the CHS versus the overall cost of the system they're used to (usually 2-x-4 or 2-x-6 wood frame). By "overall cost" we mean the total cost of the materials and the labor to install it, plus the savings or costs from any side effects the system might have. "Side effects" are changes in the costs of other parts of the building, like a reduction in the insulation sub's labor and materials because the wall components include insulation already, or an increase in the finish carpenter's costs because he or she has to use more expensive fasteners to connect to an exterior concrete wall. When they compared overall costs, builders told us how much more or less they spent to build a house out of the CHS, compared to what they would have spent to build another similar house out of frame. They also told us what the features and finishes on each version of the house would be, so you can see exactly what they're comparing. These figures should help you decide whether the CHS under discussion is in the general cost ballpark you're looking for.

The second type of information builders gave us was a list of the factors that push the overall cost up or down the most. For example, having a lot of corners in the design might drive the cost of construction up a lot with some systems, but very little with others. We got rough cost estimates for these factors when we could. They should help you make a quick estimate of how much the overall cost of a CHS might go up or down for the kinds of projects you do.

The third kind of information we got was the cost per square foot. This is helpful if you want to estimate total cost for a specific house you're planning. In most cases we also got the builders to break these per-square-foot costs into the major materials and labor components. This lets you play games like estimating how much costs will change if you increase or decrease the amount of insulation or change the siding material.

To give you something to compare the square foot costs to, we've listed approximate figures for exterior 2-x-4 wood frame walls in Table 4-1. We got the figures by asking builders in different parts of the country what they paid on their last few houses in 1994. The numbers in the table are in the middle of the estimates they gave us. For 2-x-6 framing, the total cost is about $.50 per square foot higher because of the larger-dimension lumber and insulation. That brings the total 2-x-6 cost for structure and insulation to $3.05 per square foot of gross wall area, and the total including finishes to $5.25–$13.95.

Table 4-1. Approximate Costs of 2-x-4 Wood Frame Exterior Walls

Component		Cost per square foot
Structure and Insulation		
2-x-4 framing lumber		$ .75
½" plywood sheathing		.35
(not included for some stuccos)		
3½" fiberglass insulation		.15
Glue and nails		.15
Labor		1.05
Subtotal, structure and insulation		$2.45
Finishes		
Exterior siding (material and labor)		$ .80–9.50
vinyl	$ .80	
aluminum	1.20	
stucco and paint	2.50[1]	
cedar clapboard and paint	3.40	
brick veneer	9.50	
Interior sheetrock (material and labor)		.80
Interior paint (material and labor)		.70
Subtotal, finishes		$2.30–11.00
Total, structure and insulation and finishes		$4.75–13.45
With vinyl siding	$ 4.75	
With aluminum siding	5.15	
With painted stucco siding	6.10	
With painted clapboard siding	7.35	
With brick veneer	13.45	

[1] Many modern stucco systems over frame replace the plywood or OSB sheathing with a layer of backing paper or foam and a layer of wire mesh, at a total cost for backing, mesh, stucco, paint and labor of about $2.50 per square foot. But since the $.35 for sheathing is saved, we subtract it out before getting the total cost of 2-x-4 frame and stucco.

Two quick warnings about costs per square foot. First of all, when the trades and manufacturers give the cost per square foot of anything that goes into an exterior wall (like framing labor, lumber, paint, blocks, masonry labor, reinforcing, ready-mix concrete, form work, insulation, and so on), 85 percent of the time they mean the cost per square foot of gross wall area. They aren't using the floor area of the house for their square footage number. And they don't subtract the openings from the wall area because, they say, the cost of working around an opening is about the same as the cost to build through it. To be consistent with the trades, we do it the same way. The frame costs we list in Table 4-1 are per square foot of gross wall area, and so are the other wall costs "per square foot" that we give in this book. The second warning is that all costs vary from time to time and place to place. Wood frame costs are especially prone to change over time as the cost of lumber changes. And all kinds of costs change as you go from one region to the next. Your safest bet for using the numbers in this book is to get a few hard, current quotes for labor and materials in your area. If your numbers are higher than ours for a few wood products, they'll probably be higher for the other wood products as well. If yours are lower for a few concrete products, your prices are probably lower for all concrete products. The same goes for other types of materials and labor.

The fourth type of cost information we sometimes got was advice on how to send out for a bid. When you get down to building an actual CHS house, the tips we got can help you get better terms from the trades.

Learning costs

The total cost of using any new product is higher at first. The builder spends time figuring it out, the workers spend time learning how to use it, they install it more slowly until they build up some skill, and the job has more waste. Before choosing a CHS, you need to factor in this up-front investment. Under the "Learning costs" heading in the system descriptions we list our builders' estimates of the extra time and money they spent on their first few houses. We also tell what the causes of the biggest extra expenses were, so when you build you can plan how to minimize them.

Market appeal

The frame builders we interviewed brought up eight features of exterior walls that are important to home buyers. They wanted to know how each CHS rated on each of these eight features to decide whether a house built of the system would attract shoppers in their area, get them to buy, and make them happy with their purchase.

Buyer reaction

The "buyer reaction" to a CHS is the overall opinion that home shoppers had of the system when they saw or heard about it. In other words, did people like the system? Did they understand it? Were they impressed when they heard about the benefits? Did they buy it? We asked builders, salespeople, and real-estate agents how the use of a CHS affected the buyers' level of interest in a house and why.

Buyers' reactions to a new system are related to the system's other market appeal features: exterior and interior finishes, design flexibility, R-value and energy efficiency, thermal mass, water resistance, maintenance required, and disaster resistance. It's true that some buyers have a gut preference for one material or another. For example, builders in Arizona and Florida told us that masonry has a higher perceived value there than frame, while in parts of New England we heard the opposite. But most people seem to base their final decision as much on how the finished system looks and what it can do for them at the price. The "Buyer reaction" section discusses how and how much all of these things influenced buyers considering a particular CHS.

Exterior and interior finishes

We learned that the finishes are the biggest single influence on people's reaction to a wall system. For most buyers, if the external appearance is unappealing, little else can move them to buy.

Like almost any wall system, the CHSs can take any popular exterior siding and interior wall finish. The real question is, at what cost? Just as stucco goes less expensively onto concrete than frame, and clapboard goes more easily onto frame than concrete, the different sidings and interior materials have different costs on different CHSs. In the "Exterior and interior finishes" section for each CHS, we cover what finishes have been used with it and what their costs and logistics are.

One siding that we need to present some facts on is stucco. It's true that only some varieties of stucco are "new" themselves. But stucco is widely used with the CHSs, often in novel ways that influence costs, appearance, water resistance, and maintenance requirements.

The word "stucco" actually refers to a range of different material mixes used as troweled sidings. They can go over many different wall surfaces, including over limitless details built up out of foam or mesh-covered frame.

Nearly all of the stuccoes contain portland cement, sand, water, and lime. The biggest difference is the amount of acrylic in the mix. Acrylic is a kind of polymer, or plastic. It's also in latex paint, and as with paint, the acrylic gives stucco some flex and water resistance—at a higher cost. There are hundreds of different stuccoes with different proportions of the key ingredients. But the trades and manufacturers put them into three broad categories, based on how much acrylic they contain. Table 4-2 lists the properties of each one.

Table 4-2. Varieties of Stucco

| Variety (Abbreviation) | Flexibility | Approximate Cost per Square Foot[1] | | Approximate Water Vapor Permeance[2] (perms) |
		on Concrete	on Frame	
Portland Cement (PC)	Low	$1.50	$2.50	30
Polymer-Modified (PM)	Medium	3.00	4.25	11
Polymer-Based (PB)	High	3.50	5.00	15

[1] The cost to the builder of all labor and materials to put a complete siding of the material onto a typical house. Includes the cost of painting (for PC and PM stucco) or premixed coloring (for PB stucco).

[2] This assumes an average coat thickness. For each type of stucco, the permeance is lower the thicker the coat.

Portland cement stucco is also called PC or traditional stucco. It has no acrylic. It's the least expensive of the bunch, and the ingredients come premixed or off the shelf from dozens of manufacturers. It's also hard and durable. To cure properly it needs to be kept wet for a couple of days after being applied. This is not always done in practice. If it isn't, the result is usually some hairline cracking on the surface. PC stucco is also prone to some cracking if the foundation settles. But cracks are usually only an aesthetic concern. They worry people more than they cause real problems. Some builders we met simply planned to send a crew back to their houses six months after completion to patch the cracks with an elastomeric (flexible) grout and paint over. They claimed that the cost of doing this was small, and PC stucco was still cheaper than the other two types.

Polymer-modified stucco is also called PM or hard-coat stucco. It's PC stucco with a little acrylic mixed in. Like PC, it's available from countless manufacturers. Its big advantage is that it has enough give to crack less than PC does. You have to weigh this against its higher cost.

Polymer-based stucco is also called PB or soft-coat. It runs about 50 percent acrylic. It's the most resistant to cracking and the most expensive. It's also softer than the others, but this might not be a consideration if it's put on top of something hard like concrete. More often than the others, it's manufactured by national companies and is referred to with brand names (like Dryvit, Sto, and Acrocrete).

With all the stuccoes, it's possible to mix pigment into the top coat so that you don't have to paint. If you do this, the owners won't need to repaint. Unless there's wall damage, the color lasts indefinitely. But the pigment adds cost, too, and unless mixing of all ingredients is precise, the color can vary across batches. Most users of PC stucco we asked preferred painting. The PB and sometimes the PM stuccoes we saw were precolored. The larger manufacturers of theses stuccoes have standard colors that are precisely premixed at the plant.

The type and cost of stucco used depends a lot on the climate. PC is the most common choice in the South, but in cold climates builders and installers said they frequently chose a PM or PB instead because these withstand freeze-thaw cycles better without cracking. And when PC was used in cold areas, it was usually with extra care at extra cost. In some of the southernmost states, we got quotes of PC stucco and paint on block for less than $1.00 per square foot. In the northernmost states, we got some for more than $3.00.

Design flexibility

One sharp difference among wall systems is the ease with which you can form them into different sizes and shapes of structures. As we've discussed before, you can make any system do just about anything with enough money and monkeying. But some systems can form basements, go up multiple floors, or form non-90-degree angles, bay windows, curved walls, and irregular openings much more easily and inexpensively than others can. For each system we asked the builders which of these things they'd tried and what their costs and level of effort were. As you'll see, some of the new systems are much more flexible than traditional concrete—and some even more than frame.

R-value and energy efficiency

A lot of builders looking at concrete wall systems are particularly concerned about R-values. They remember that the old concrete systems had low ones, and it was difficult and expensive to add insulation.

A lot of the new CHSs obliterate this problem. They do it in two ways. First, they achieve high R-values by using rigid foams and clever design tricks to form thick layers of nearly uninterrupted insulation. Second, they take advantage of the high mass of concrete to even out temperature swings. It turns out that this so-called "thermal mass effect" lets concrete homes consume less heating and cooling than frame houses that have the same R-value. For the same level of insulation, you get a lower fuel bill with concrete. That's why a lot of people now talk about the total energy efficiency of a house. R-value is only one of the major factors that determines the cost of heating and cooling. And if that weren't enough, the thermal mass effect has another benefit. Because it evens out temperature swings, it reduces peak heating and cooling loads. So you can put smaller HVAC equipment into a house, saving hundreds of dollars up front.

If you want to pick your wall system well, you need to be able to compare energy efficiencies and explain them to buyers. But it's not easy. Different manufacturers calculate their numbers in different ways that produce very different results. So you have to know what all the calculations are and what they mean.

R-value is a measure of what engineers call thermal resistance. The thermal resistance of a material (or wall) is how fast it lets heat pass through. Everything but a vacuum lets heat through. It's just a question of how fast. But some materials "hold heat in" (or out, if you're air conditioning) better than others. The more slowly the heat passes through, the higher the material's thermal resistance is said to be, and the higher the R-value we give it. For those interested in technical details, if one square foot of a material lets one Btu (British thermal unit, a measure of heat) pass each hour from a hot side at 71 degrees to a cold side at 70 degrees, it has an R-value of 1. If it lets only half a Btu through, it has an R-value of 2, and so on. As we'll see, a one-inch sheet of a typical foam has an R-value of 5. Each square foot lets one-fifth of a Btu through each hour when the temperature difference between sides is one degree. Layering materials adds to the total R-value. So 2-inch sheets of foam, or two 1-inch sheets on top of each other, have an R-value of 10. Only one-tenth of a Btu gets through.

Table 4-3 contains R-values for typical insulation materials used in wall systems. One of the most commonly used materials in the new CHSs is expanded polystyrene (EPS). It's a plastic foam that comes in two varieties. Molded EPS is the same material used to make vending machine coffee cups. It's molded into a wide variety of shapes out of tiny beads that give it a sort of "grain." It's so widely manufactured that it usually has no brand name, but the board stock used for building insulation sometimes goes by the nickname "bead board." Extruded EPS is the material used to make backing trays for fresh meat in grocery stores. Because it's extruded, it has no grain or seams. It's nearly continuous.

Table 4-3. Properties of Insulation Materials

Material	Approximate R-Value Dry	Wet[1]	Approximate Price per Square Foot[2]	Approximate Water Vapor Permeance (perms)
Fiberglass				
3½" batts	11	NA[3]	$ .16	NA[4]
5½" batts	19	NA	.26	NA
Molded EPS Bead				
1" board	5	3.6	.18	2.0–5.8[5]
2" board	10	7.2	.36	1.0–2.9
3" board	15	10.8	.54	0.5–1.5
Extruded EPS				
1" board	5	4.8	.36	1.2
2" board	10	9.6	.72	0.6
3" board	15	14.4	1.08	0.3
Polyisocyanurate				
15⁄16" board	7.2	4.3	.43	4[6]
2" board	14.4	8.6	.81	2[6]
3" board	21.6	12.9	1.20	1[6]

[1] These R-values are taken from boards of insulation that have been buried below grade along the wall of a foundation for 18 months.
[2] This is the retail price, not including any quantity discounts.
[3] Fiberglass is not normally tested for R-value when wet, but it would probably rate very low.
[4] Fiberglass in not normally tested for water vapor permeance, but it would probably rate very high.
[5] The water vapor permeance of molded EPS bead depends heavily on how it's manufactured. The tighter the beads, the less vapor gets through and the lower the permeance.
[6] This assumes the board is unfaced. Facing reduces the permeance.

The most common brands for building insulation are Dow Styrofoam (which is blue) and UC Industries' FoamulaR (pink).

A couple of warnings about R-values. First of all, the same type of insulation can have a different thermal resistance, depending on how it's manufactured. Most insulations can be made with a lot of small air pockets (in which case they have a high density) or fewer, larger ones (in which case they have a low density). The higher-density versions of the insulations have higher R-values. You can actually buy low-density 1-inch EPS board that has an R-value of only 3. But most that's sold is R-5 per inch, so that's how we list it here. You can also get 3½-inch batts of premium fiberglass that measure R-14, but they're the less common variety. Second, R-values drop when insulation gets wet. This can happen if it's used below grade or isn't protected from weather. But sellers list the R-value of their insulation when it's dry. To give you an idea of how much the R-values fall, Table 4-3 also includes the values found for different types of foam board buried below grade along a foundation. Because of its especially high water resistance, some builders favor extruded EPS for below-grade and exposed applications, despite its relatively high price.

Measuring the R-value of a whole wall is more complicated. This is because walls aren't a slab of all one material. They're a combination of different materials that cut across one another in irregular ways. Depending on how many of these pathways you take into account, you can get very different R-values for the same wall.

Table 4-4 lists the major ways to determine a wall's R-value, from simplest (and probably least accurate) to most complex. The simplest is what we call the insulation R-value. It's just the R-value of the insulation in the wall. Builders commonly call a 2-x-4 stud wall filled with fiberglass batts an "R-11 wall" because that's the R-value of the batts.

Table 4-4. Methods of Determining the R-Value of a Wall.

Method	Factors Considered	R-value Assigned to a Stud Wall[1]
Insulation R-value	R-value of the insulation	11
Nominal wall R-value	R-value of the insulation R-value of other layers	13
Parallel path R-value	R-value of the insulation R-value of other layers Insulation breaks	10
Isothermal planes R-value	R-value of the insulation R-value of other layers	9.7

Table 4-4. Continued

Method	Factors Considered	R-value Assigned to a Stud Wall[1]
	Insulation breaks	
	Lateral heat movement	
Guarded-hot-box R-value	R-value of the insulation	9
	R-value of other layers	
	Insulation breaks	
	Lateral heat movement	
	Manufacturing and assembly variations	
Mass-corrected R-value	Wall R-value as determined by one of the above methods	Variable[2]
	Thermal mass	

[1] This assumes a 4-inch wall with softwood studs 18 inches on center, ½" sheetrock inside, fiberglass batt insulation in the cavities, and ½" plywood sheathing and red cedar clapboard outside.

[2] This varies depending on local climate. However, because of the low thermal mass of a frame wall, its mass-corrected R-value will usually be little different from a method that does not take thermal mass into account.

But the insulation R-value neglects the other layers of the wall. If you add the R-values of the sheetrock, sheathing, and siding to the R of the insulation in the stud wall, you get what we call the nominal wall R-value. It takes into account the entire wall, but it's still "nominal" in the sense that it neglects other important details. With plywood sheathing and clapboard siding, the nominal wall R for 2-x-4 frame is about 13.

The nominal wall R-value ignores that there might be breaks in the insulation. In a typical 2-x-4 wall, the studs, plates, and other framing members actually make up about a quarter of the total area. Lumber doesn't insulate as well as fiberglass does, so the wall as a whole insulates less well than it would if it had no studs—if it had only unbroken insulation. Engineers correct for this extra heat loss with a formula that averages the R-values of all straight paths through the wall. They often refer to it as the "parallel path method." So we refer to R-values calculated this way as parallel path R-values. The formula estimates the R-value of the stud wall described above at about 10. This is the simplest calculation that engineers would normally accept in their work.

But heat can also go sideways for part of the journey through the wall. This is important in walls with breaks in the insulation (like the studs) and different layers of materials. In a 2-x-4 wall, heat on the inside moves into the sheetrock. The sheetrock is a poor insulator, so the heat moves readily sideways and "leaks" into the cold spots in the next layer—the studs. Thus the heat passes through a little faster than the parallel path formula assumes. To account for lateral heat movement, engineers use a more complicated formula they call the "series parallel" or "isothermal planes" method. We therefore call the resulting R-values isothermal planes R-values. Using this formula on the 2-x-4 wall described above yields a value of 9.7. This might not seem like a big change from the parallel path estimate of 10. But the difference can be larger for the CHSs covered in this book. They often have alternating layers of concrete and foam, just the sort of configuration in which lateral heat flows can be large.

But we still haven't corrected for variations in manufacturing and assembly. When people calculate the R-value of a stud wall, for example, they assume that the fiberglass completely fills the spaces between the studs. But in practice there are usually gaps. These increase the size of the insulation breaks and lower the thermal resistance of the wall. To correct for this, another formula wouldn't be enough because no formula could include all the separate imperfections in a wall. The most widely respected method is a test called the *guarded-hot-box test*. Lab technicians build a wall down the middle of a special box. They keep one side of the box warm and one cold, and they measure how fast heat flows from the "hot" side to the cold. It's generally considered an extremely accurate method of determining R-value. And especially for walls that get put up with inconsistent materials in the field, it can show some very different results. The director of a testing laboratory told us his crew had built a 2-x-4 stud wall and tested it in their guarded hot box at R-8. Take this number with caution because he was probably rounding off, and he didn't give all the details of the size of the gaps or the finishes used. Some architects told us they assume that an average 4-inch stud wall will be R-9. We call R-values measured in a guarded hot box *guarded-hot-box R-values*.

But even the guarded-hot-box test has limitations. As an engineer will tell you, it's "static." It assumes things aren't changing. In the test, each side of the box is maintained at a constant temperature. But in the real world, the air temperature fluctuates. And in these conditions, the thermal mass effect reduces the total heat passed through the walls of a house. In some concrete houses the fuel savings resulting from the effect have been calculated at more than 25 percent.

The thermal mass of a wall is how much heat it holds. Each material has its own thermal mass, just as it has its own thermal resistance. People made use of this principle a century ago when they heated stones on the hearth before going to bed and put them under the covers to keep their feet warm. It took an hour or so for enough heat to enter the stones to get their temperature up. But then they contained enough heat to stay warm for an hour or two. If our forefathers had put a bag of cotton on the hearth instead, it would have reached maximum temperature faster. But it would have cooled off within a few minutes because cotton holds much less heat at the same temperature.

The impact of thermal mass on heating and cooling is dramatically demonstrated by the old adobe homes of the southwestern deserts. They had thick walls of a heavy material (actually a form of mud) and no insulation. The hot sun often brought the daytime outdoor temperature to more than 100. Yet the adobe home would stay relatively cool. The sunlight and warm air striking it would constantly pass heat to it, but it had such a great thermal mass that it would absorb this heat without its own temperature rising greatly. By evening it would be warm. But then the sun would be gone, and the temperature could drop to the 50s overnight. The warm adobe would pass its heat back to the air, and in the process keep the inside temperature tolerable until morning. The next day the process would start all over. Measurements on adobe homes in Arizona and New Mexico showed some interior temperatures that centered in a narrow band around the mid-70s, despite the absence of any insulation or HVAC equipment.

In the same climate, an uninsulated, unheated frame house would be uncomfortably warm during the day and cold at night. It would simply heat up too quickly and cool down too fast because its materials hold so little heat.

Even in insulated houses, thermal mass makes a substantial difference. As you might expect, the difference is greatest in moderate climates where the outside air temperature crosses over the comfort range frequently. Measurements indicate that some houses in southern Florida built of block with R-15 walls only use as much air conditioning as a frame house with R-23 insulation. In other words, the block houses needed about two-thirds as much cooling over the course of a year as a comparably insulated frame house needed.

In cold climates the temperature crosses the comfort zone less often, so the energy-saving effect of the thermal mass is less. Calculations suggest that a house of block with R-15 insulation in Bismark, North Dakota would save about 5 percent of its heating bill over a similar frame house.

There's no one way to calculate the effect of thermal mass that's accepted by everyone. There are laboratory tests, computer models, and formulas. However, many of these methods produce a so-called mass-corrected R-value for a wall system. This is the R-value you would have to add to a low-mass wall to get the same energy efficiency as the wall being measured. In the example mentioned previously, the walls of the block house in Florida would be assigned a mass-corrected R-value of 23. They don't have that much insulation, but they perform like a frame house that does.

It's important to remember that the mass-corrected R-value of a wall system depends on where it is. The same wall that had a mass-corrected value of 23 in Florida has one of 16 in North Dakota. So to make use of a mass-corrected R-value, you have to know where it was calculated for. It's also important to remember that the mass-corrected R-value starts with one of the other R-values, then adjusts it. For accuracy's sake you'd probably prefer a mass-corrected R-value that starts with the guarded-hot-box or isothermal-planes R, but that's not always the case. If you can find out which R-value was used as the starting point, you can judge the accuracy of the mass-corrected version.

So now we know the different types of R-value. Which one do building departments use? Which one are buyers interested in?

Most building departments we talked with just asked that the R-value of the insulation meet some minimum limit. So technically they were using the insulation R-value. Some would let builders add the R-values of the other layers, which amounts to using the nominal wall R-value. The way most codes are written, the building departments are allowed to use more complex methods. But few seem to. With a preassembled block or wall system, they might ask for spec sheets that list an R-value to make sure it's above their minimum. But they never asked how that R-value was determined. They generally demanded that the advertised R-value of the new wall system be at least as high as the value set for frame, even though the wall system's R might have been determined with some much more demanding method like the guarded hot box.

The simple methods of the building departments might sound odd after all the evidence that the insulation R-value overstates the true thermal resistance and that different R-value calculations are not comparable. But given the departments' limited time and resources, these practices might be good enough. For most of the new CHSs, the methods that the building officials use aren't crucial because the system has more than enough insulation for even northern climates. But to play it safe, it might be worth running your favorite system by the building department early to see how they rate its R-value. If they're way off, the system manufacturer will usually help educate them.

An increasing number of codes are now allowing builders to take thermal mass into account. An advantage of this for builders is that they could meet code with less insulation. For example, in a town requiring an R-13 wall, the builder might get approval with only enough insulation to produce, say, an R-11 wall. On a mass-corrected basis, the wall would get counted as R-13 in some areas. But because there's no generally accepted way to determine mass-corrected R-values, the codes don't use them to determine how much of a break to give on insulation. Instead, they have tables of numbers that go by names like "multiplier factors" or "correction factors." You and the building officials use these tables to determine how far upward to adjust your R-value. The tables and how they work differ from code to code, but usually they take two or three of the following things into account:

The amount of thermal mass in the wall Obviously, the greater the thermal mass, the greater the effect and the more of a correction you should get. To measure thermal mass, some codes just use the pounds of concrete per square foot of wall area. For example, a standard 8-inch block wall usually weighs 35–40 pounds per square foot. A solid 4-inch concrete wall is about 45–50. Some other codes use a measure of how much heat the wall holds—Btu/square foot/degree Fahrenheit. The 8-inch block wall is usually about 8 Btu/sf/degree. This means that every time the wall falls by one degree, each square foot has given off 8 Btus of heat. A solid 4-inch concrete wall is about 10–11 Btu/sf/degree. Because we don't know which of these two measures your local code uses, we try to give both figures for each CHS when we describe it.

Climate Nearly all codes give more credit for houses built in warm climates than cold ones. The typical way they measure climate is with degree-days. Simply put, the fewer the degree-days in your region, the warmer the climate. Local code officials should be able to tell you the degree-days in their area. Once you have it, you can look on the right line of the table.

Placement of the thermal mass Engineers' studies show that the thermal mass effect is greatest if the insulation is on the outside of the mass, second greatest if the insulation is in the middle of the mass, and least if the insulation is on the inside of the mass. So you get the most credit if the insulation goes outside the concrete, and the least if it's inside. The difference is small, but some code tables require you to look under a different section, depending on which arrangement your system has.

Interestingly, we found that a lot of the CHS builders don't apply for the insulation credit even though their systems have more than enough thermal mass. This is partly because the credit is new. Some local codes don't allow it yet. Others do, but the local officials aren't familiar enough with it to help builders figure it out. But in addition, a lot of builders didn't really need the credit. Most of the CHSs have R-values that are plenty high for almost any local code. And the builders wanted to offer greater energy efficiency, not the minimum amount of insulation the code allows. But even for these builders, the thermal mass provisions in the code were still useful. They provided proof to buyers that the thermal mass effect is for real, and the tables gave them a rough idea of how much of a difference the mass makes in their area. But you might want to bear in mind that if one of the CHSs is below the R-value requirement of your area, and if local code allows counting the thermal mass, using the tables might put the system over the limit.

As for buyers, which R-value they want to know depends a lot on the individual. For most, just telling them that their fuel bill will be lower than their neighbors' is enough. More knowledgeable buyers with an interest in energy will be impressed if you understand and can explain the energy efficiency of the CHS you're offering.

Thermal mass

Thermal mass can do more than just lower standard heating and cooling costs. It's also crucial for passive solar and load-leveling homes. In solar homes it holds the heat that comes through south-facing windows. In load-leveling homes, it holds coolness generated at night to keep the living space comfortable during the day.

For both of these uses, it's the thermal mass inside the insulation that counts. The objective is to bring heat or coolness into the house and keep it there until other times when it's needed. The mass outside doesn't contribute to this.

How much thermal mass contributes to solar heating or load leveling depends on so many factors that it takes an engineering study to estimate its impact on any particular house. But we can give you a feel for it with some examples.

We saw calculations for a house with walls of ordinary 8-inch block, 3-inch foam attached to the outside, double-pane windows, and R-35 attic insulation. They predicted that in 32-degree weather with no sunshine (as at night), the temperature inside the house would drop about 7 degrees (from 72 to 65) over 12 hours. So the house would make it through the night more or less comfortably without heat.

Near Orlando, Florida, electrical utilities and the Florida Concrete and Products Association have sponsored the construction of what they call the Off-Peak Thermal Mass (OPTM) house. Its walls are block filled with grout, forming an 8-inch solid concrete wall. The air conditioning is on a timer that shuts it off every day at noon and only allows it to turn on again at 9:00 p.m. According to the residents, on even hot summer days the inside temperature only rises about 5 degrees (from, say, 70 to 75) during that time. So the sponsors hoped, during the hottest part of the day the block absorbs so much heat without rising much in temperature that the living space stays in the comfort range.

The best way we know to measure a wall's ability to store solar heat and off-peak AC is with the same Btu/sf/degree Fahrenheit measure of thermal mass that some codes use to figure insulation credit. If you increase or decrease the Btu/sf/degree on the inside of a wall (by, say, adding or taking away concrete inside the insulation envelope), the amount of heat or coolness that it stores goes up or down proportionately. For the ordinary 8-inch block wall described previously, the inside thermal mass is about 8 Btu/sf/degree. Technically, that means that every time the walls absorb 8 Btus for each square foot, they rise 1 degree. But you can just think of it as the amount of storage necessary to get a well-insulated house through the night (falling from 72 to 65) without heat in freezing weather. If the wall were solid 8-inch concrete, the thermal mass would more than double to 21 Btu/sf/degree. In that case, it would last over two-and-a-half times as long in freezing weather before dropping to 66 degrees.

The OPTM house described previously has a mass of about 21 Btu/sf/degree. Technically, this means that every time each square foot of the wall absorbs 21 Btu, it rises 1 degree, and every time each square foot gives off 21 Btu, it falls 1 degree. But you can just think of 21 Btu/sf/degree as the amount of storage necessary to get a well-insulated house in Florida through nine hours of heat without air conditioning. If the walls had been 4 inches, they would have had 10.5 Btu/sf/degree, and the house might only have gotten through four to five hours without a need for AC.

For purposes of comparison, most wood frame walls have about ½ Btu/sf/degree. That's the thermal mass of the sheetrock, which is about all that's inside of the insulation. With that amount, a well-insulated house in freezing weather should fall from 72 to 65 degrees in about 45 minutes. On a summer day in Florida, it should last about 30 minutes without a need for air conditioning.

The different CHSs vary widely in their inside thermal mass. Under the system descriptions, we give the numbers when we have them, plus the stories we've heard about houses designed for solar heating or load leveling.

Water resistance

Like wood, concrete is a porous material. It's filled with tiny channels that water can move through. If rain (above grade) or groundwater (below grade) presses against the outside of a concrete wall, it enters these channels. If the amount of water is small, it will probably drain down and back out before it passes all the way through the wall. But at the high pressures of a severe rain storm or standing groundwater, the water can pass all the way to the interior.

Builders expressed concern about the CHSs' resistance to water. Probably they remember the leaky concrete systems of the past. Every new CHS we looked at has at least one feature to prevent water penetration. To give you a measure of how well they work, we report the available water resistance data for each system.

Probably the best measure of a wall's water resistance is the driven rain test. Lab technicians set up the wall, then spray water against one side at a rate equivalent to a downpour driven by high winds. They then measure how much of the inside wall surface water seeps through. In this test, a wall of 8-inch block with ordinary latex paint shows water over about 60% of its interior wall surface after eight hours.

Unfortunately, manufacturers aren't required to perform this test on their products, and most of them don't. So the data is available for only a few of the new systems. The test is also done only on masonry products, so there are no figures for solid concrete or frame walls.

An alternative measure of water resistance is water vapor permeance. This is how fast water vapor (not rain or groundwater) goes through a material. It's actually a second-best measure, but the numbers are more often available because many materials have been tested for water vapor permeance. Some people assume that if you have a continuous layer of a low-permeance material in a wall, you're well protected against water penetration. If the layer has a lot of breaks in it, or if none of the layers of the wall are of low vapor permeance, they have less confidence in the wall.

The permeance test is done in either of two ways, called the *wet cup* and *dry cup* methods. In the wet cup method, technicians put a wall section or slab of material over a container of water in a dry room and measure how fast water vapor seeps through. If the material allows water through at the rate of one gram per square foot of material per hour, it has a permeability of one perm. The dry cup method is the same except that the cup has no water, the room is humid, and the technicians measure how fast vapor goes into the cup. The dry-cup reading is usually about half of the wet-cup reading for the same material. We cite the wet-cup readings, which are a little more commonly used. To give you a basis of comparison, a plastic vapor barrier with no punctures tests at about one perm by the dry-cup method. That's considered a very low rate of permeance.

Although each of the CHSs has its own features for resisting water penetration, there are three types of barriers that many of them use. They're foam, stucco, and paint.

As you can see in Table 4-3, some of the foam insulations have very low water vapor permeance. None of the CHSs rely on foam as the primary line of defense, but to some builders a low-perm foam in a continuous barrier in a wall looks like it would form a big obstacle to water.

The stuccoes don't have vapor permeances that are quite so low, as listed in Table 4-2. But they're much denser materials than structural concrete. This makes it more difficult for water in its liquid form to pass through. And the PM and PB stuccoes contain acrylic, which reduces water transmission sharply. Several systems do rely on stucco for some of their defense against water, and as you'll see, it generally seems to work.

Paint is a commonly used barrier for all wall systems. Ordinary latex resists water penetration, whether it goes on wood or concrete. Depending on the exact formula, a two-coat layer of latex has a permeance of about 2–6 perms, meaning that it allows about 2 to 6 times as much vapor through as a vapor barrier does. A few of the CHSs rely on latex or a special concrete paint, but usually in combination with some form of stucco for a double barrier. Some CHS builders we spoke with went to the extra expense of an elastomeric paint. These special paints go on thick and have a lot of give. Once dry, they can be stretched to over twice their original dimensions without tearing. They keep water out effectively. Their vapor permeance is usually below one-half. And, incidentally, they can stretch enough to cover most hairline cracks that open in the wall underneath. However, they run about twice the materials cost of conventional exterior paint, around $.20 per square foot.

Most good painting contractors can be counted on to come up with a suitable water-resistant paint for a concrete or stucco surface. They'll have to go to specialty supply houses for the more exotic ones like elastomerics.

Block walls covered with stucco and paint have been subjected to the driven rain test. In most cases little or none of the interior surface is wet after eight hours.

If you use a system that relies on paint alone for its water resistance, remember that repainting becomes crucial. Just as with wood-sided frame walls, when the paint starts to go, so does the water protection.

Maintenance required

The most common causes of exterior wall maintenance that homeowners talk about are termites, rotting, and paint peeling or fading. The first two rarely afflict concrete walls. So there's not much need to talk about them here, except to say that virtually all of the new CHSs experience structural damage much less often from pests and water than frame does.

How often a wall has to be repainted actually depends more on the siding material than the structural material. But the CHSs have some form of concrete surface more often than frame houses do. So it's useful to compare painting requirements for different sidings.

According to manufacturers' recommendations, concrete and cement-based surfaces should be repainted about every six years. For wood and wood-based sidings, the recommendation is every four years. The experience of homeowners bears this out. Paint tends to fade or peel after these times. So cement-based surfaces require painting only two-thirds as often. Today, hiring a crew to paint the exterior of a two-story, 2000-square-foot (living space) house runs around $2000. Having to do this every six years instead of every four works out to spending $333 per year instead of $500.

Disaster resistance

Statistics from the American Insurance Association show that the most prevalent source of disaster damage to homes is fire. Wind is second, and earthquake is third. Historically, the damage and loss of life from fires has been several times that from wind, which is itself several times the loss from earthquake.

Different wall systems have sharply different abilities to withstand disasters. The most common way to measure resistance to fire is with the fire wall test (known to engineers as "ASTM E119"). There are some variations on the test, but in most cases technicians in a lab burn a gas fire at a controlled temperature on one side of a wall until the cool side overheats past certain temperature limits. If the wall survives intact, it gets a fire wall rating equal to the length of time it was subjected to the flames—1 hour, 1½ hours, or whatever the time was. If instead the wall fails structurally during the heating, it gets a fire resistance rating instead. That indicates that the wall might prevent fire from spreading for the length of time it stayed below the temperature limits, but is less likely to be intact at the end.

Most 8-inch block walls are fire wall rated at two hours or more. After that time in the test, they overheat on the cool side. They rarely fail structurally. A 2-x-4 frame wall with sheetrock on one side and wood siding on the other is generally rated at one hour for fire resistance. After that the wall fails. Under the system descriptions, we give the results of fire tests on the CHSs, plus any stories about how they survived actual fires.

Insurance statistics confirm what the tests and common sense tell us: concrete walls have a higher fire survival rate than wood frame. Particularly impressive for exterior walls is their ability to remain standing through a fire, rather than collapsing. In most areas of the country, occupants get a reduction in the fire portion of their home insurance if the house has concrete walls. It amounts to about $40–50 per year for the typical home.

The most common measure of a house's ability to withstand wind is the wind speed rating. This is an estimate of the fastest wind that the structure can take without breaking apart. For a specific house, engineers calculate these wind ratings from laboratory tests that give the strength of each part of the building (for example, the strength of the studs and the strength of the nail connections on a frame wall).

Most building codes require that houses be built with components and designs that have been shown to meet some minimum wind rating. The highest requirement in the United States that we're aware of is 130 mph in various coastal areas. Inland, 80 mph is a more typical number.

Like a lot of component manufacturers, the makers of CHSs often have engineers calculate wind speed ratings for typical wall designs. We present these numbers when we have them, as well as stories about how the systems fared in actual storms.

Some general information on the performance of concrete walls in high winds comes from three recent studies of houses that were in the path of Hurricane Andrew. All concluded that the concrete houses (mostly block) suffered less damage than the frame. One was done by the NAHB Research Center, one by a team headed by a Florida State University engineering professor, and one by a builder who wrote up his personal observations in the December 1992 issue of *Fine Homebuilding*. Most of the damage in that storm was actually to the roofs. On concrete walls, roofs are often attached with metal straps (nicknamed "hurricane straps") anchored directly into the concrete at the top of the wall. Compared with the simple nails that connect roof to wall in many frame houses, these straps are difficult to pull out. Somewhat less damage in Andrew occurred to the houses' walls. The wind had pulled some walls apart, and flying debris had punctured some. The wall damage was sharply lower for the block houses. Their weight and the standard use of steel reinforcing bars make block and other concrete walls highly resistant to separation. When they have continuous exterior surfaces of concrete, they are also resistant to punctures.

A structure's resistance to earthquakes can be determined by a laboratory test called a *shake table test*. A complete wall is mounted on a large platform that moves to simulate the vibrations of an earthquake. But it's so expensive that very few manufacturers have one done. In most cases, seismic resistance is estimated the same way wind resistance is—by engineers using formulas that make the best guess of what will happen to a particular wall in an earthquake, based on past experience and laboratory tests of the components. A lot of CHS manufacturers have these calculations done, and we report the results and any stories about the systems' experiences in actual quakes.

Most manufacturers give the level of earthquake resistance of their products by telling you the "Zone" they're approved for. Testing and code organizations have divided a map of the United States into higher-risk and lower-risk seismic areas. The lowest-risk parts of the country they call Zone 0 seismic areas. These are the areas that rarely experience a quake, and only mild ones when they do. A lot of the Plains States are rated Zone 0. Regions like southern California, where earthquakes are relatively frequent and strong, get a Zone 4 rating. Areas with a risk level somewhere in between are classified as Zone 1, 2, or 3. Wall systems that are estimated to be capable of surviving only the mildest of earthquakes are rated Zone 0, and those that can survive a typical Zone 4 quake are said to "be approved for Zone 4." Not all local codes require that buildings be able to withstand quakes of the level specified by their area's zone rating. But more and more do. In any case, the Zone rating for a CHS is a handy indicator of how earthquake resistant it is.

Homes of traditional concrete systems appear not to have fared as well in earthquakes as wood frame. We might expect this from the stiff nature of the material and the small amounts of steel reinforcing used in the past. In high seismic areas like California, insurance companies have charged higher premiums on policies for block homes, just as the fire premium is higher for frame in many areas. But as you'll see when you read about the new CHSs, some can match or outperform the traditional wall systems (concrete or lumber) with little or no additional reinforcing.

When reading the stories about actual earthquakes, bear in mind that the "Richter scale" that's used to measure the severity of a quake is an exponential scale. That means a quake measuring 2 on the Richter scale is 10 times as severe as one measuring Richter 1. A quake measuring Richter 3 is 10 times as strong as one measuring Richter 2, and so on. The kind of quake Californians call "the big one"—a quake severe enough to cause 7 billion dollars of property damage if it were centered near Los Angeles—would, experts say, measure 8 on the Richter scale.

Availability

Regardless of how attractive a system is on other grounds, it's useless if you can't get hold of the parts and people to install it. Like many new products, some of the CHSs are more readily available in some areas of the country than others. The same is true of the workers qualified to erect them.

Product availability

The manufacturers of all of the CHSs have lined up plants to make the special parts and materials of their systems. Sometimes these plants can get products economically to almost any point in the United States. But in some cases they're far enough apart that delivery time and shipping charges are high to some locations. In extreme cases, the added time and cost can make a system impractical for use in some areas.

When you read about the product availability of a particular CHS, use some caution. Many of the CHS manufacturers are setting up new plants and expanding their shipping schedules. We report their expansion plans when we know them. But for any system that interests you, it might be worthwhile to call the manufacturer about availability before making any final decisions. The company might have changed its expansion plans. A system you'd love to use might unexpectedly be for sale near you in the near future. Or it might unexpectedly no longer be available there.

Labor availability

Some of the new CHSs involve unconventional tasks, such as surface bonding or foam form construction. Depending on where you are, it might be easier or harder to get qualified crews.

The general options for getting a crew with specialized skills are to track down a crew that's done the task before, train a skilled crew in the most closely related trade, or assemble a crew and train them from scratch. But the CHS builders had some clever, original ways to handle the situation. So, under each CHS description, we relate the different things they tried and how each one worked out.

Building department approval

When someone mentions using a new product (especially one as important as a structural system), a lot of builders are convinced they'll never get it past the town building department. That's usually not true with a product that has as much testing and code documentation as most of the CHSs do. On the other hand, it's good to know what's involved in getting approval.

A lot of the uncertainty about whether the local building code will accept a new product comes from the fragmentation of the U.S. codes. In some other industrialized countries, there's one basic code for the entire nation. But in the United States, the rules change from town to town. So building officials don't always know whether a product that's accepted somewhere else will also meet local rules. It's often the builder's responsibility to provide evidence that it does.

The situation is helped somewhat by the so-called model codes. These are building codes written by a group of experts to give local building departments a starting point when they're writing their own codes. There are three major model code groups in the United States, as listed in Table 4-5. Each one writes its own model code, and each one is used mostly by building departments in one region of the country. The building departments make a few changes or write their own code almost from scratch. But in nearly all cases, they start with the model code of their region.

Table 4-5. The Major U.S. Model Code Organizations

Organization (Abbreviation)	Model Code (Nickname)	Regions of Greatest Influence
Building Officials and Code Administrators International (BOCA)	Basic Building Code (BOCA Code)	Northeast and Midwest
Southern Building Code Congress International (SBCCI)	Standard Building Code (Southern Code)	Southeast
International Conference of Building Officials (ICBO)	Uniform Building Code (ICBO Code)	Plains States and West

In this patchwork of U.S. building codes, there are four basic ways to show your building department that a new product meets the local code. We list them in Table 4-6. Which one you use depends on the system you're using.

Table 4-6. Methods of Getting Building Department Approval for a New System

Code/Testing Status of System	Requirements for Approval[1]	Useful Support Materials[2]
Variant of a system already covered by the code	Equivalence to the established system Adherence to code provisions for the established system	Manufacturer's documentation Relevant passages from the code book
Evaluation report complete	Existence of the evaluation report Adherence to the provisions of the evaluation report	Evaluation report number Copy of evaluation report
All tests complete	Existence of test results Adequacy of the tests performed	Manufacturer's documentation Copies of labs' test reports
Not all tests complete	Ability to meet all performance requirements of the code	Manufacturer's documentation Engineer's report

[1] While technically required, local officials may not ask for formal evidence of these things.
[2] Although they usually have the authority to do so, local officials may not ask for any or all of these materials.

Some new systems are considered to be just variants of traditional systems already covered in the code. For example, stay-in-place panel systems produce a pretty conventional 4- to 12-inch reinforced concrete wall, and the interior- and exterior-insulated mortared block systems have at their core a nearly standard block wall. In these cases, approval is usually easy. You just show that the structure is a conventional poured-in-place or block wall, and that it meets the usual code regulations for that system.

Some other new systems are very similar to traditional systems, but with clear differences that might affect the structure. For example, some of the cavity-insulated mortared block systems use special concrete blocks with webs (the crosspieces connecting the front and the back) that are smaller than usual. In this situation, the system manufacturers usually get the parts of the system tested as though they are the conventional products. For example, the designers of blocks with small webs have their products tested for compressive strength, dimensional stability, and water absorption. These are exactly the same tests every other block design had to go through at one time or another. By showing that their new, unusual blocks meet all the usual test requirements, the designers qualify to have their products treated like conventional blocks. When plans call for such a wall system, usually building officials either don't ask for any special paperwork, or they settle for the data in the manufacturer's product literature. If building officials still have doubts, the manufacturer will generally send copies of the test results.

For systems that are considered more than a minor variant on a conventional system, a lot of manufacturers smooth the way by getting an evaluation report (also called a research report) from one or more of the model code organizations. The manufacturers sometimes call the evaluation report on their system its "BOCA number," "SBCCI number," or "ICBO number," depending on which code body wrote it. An evaluation report on a product verifies that the staff of the code organization has reviewed the tests done and that the product appears to meet all the requirements of the code when installed correctly. The code organizations prepare these evaluation reports as a service. Instead of slogging through a lot of separate tests and engineers' reports to see whether a new system satisfies all of the model code's requirements, the building official and the builder only have to read one brief document.

In our experience, building officials accept a system with an evaluation report without much extra questioning. We've even heard of some departments that are satisfied with a report from one of the model code organizations outside of their area. For example, in one case, building officials in Georgia accepted an ICBO report for a new system even though their own code (and most others in the Southeast) was based on SBCCI's code. If you're interested in a CHS with an evaluation report, it might be worth giving a copy to your building officials to make sure they'll accept it. The manufacturers are generally happy to send out free copies.

Occasionally you run across a system that hasn't yet been tested for everything required by the local code, let alone having a complete evaluation report. It's still possible to get approval. The officials might decide that the system will meet all requirements by comparing it to similar systems. If they want more proof, they might require you to have a professional engineer analyze the product and your house design. This can be expensive and time-consuming, but fortunately, it's not usually necessary with the new CHSs.

Actually, building departments might ask for an engineer's report any time they're concerned about the soundness of a proposed house, regardless of how much testing and previous usage the materials and design have. Since Hurricane Andrew, some towns in Florida have been requiring an engineer's report on every single house, whether it's built with frame, block, or something new. But officials probably ask more often when the system is unfamiliar and there's no evaluation report or complete test results. Under the description of each CHS, we explain what kinds of approvals it has, and what builders' experiences have been when they've submitted plans to the local building department.

Required calendar time

The amount of calendar time that passes from the start of wall construction until you can begin the next task is important for trade coordination and project completion. A lot of it consists of the actual work time involved. But also included is downtime from such things as weather delays, waiting for parts delivery, and allowing materials to harden, set, or cure. We relay to you the amount of time our builders found they needed to allocate to their walls when they used a CHS. We also explain what the major components of the total time were, and what unexpected events can make it run long or short.

In a nutshell, most of the new CHSs go up in about the same time that frame does. But this is different for each system. One factor that influences the total calendar time of a lot of the CHSs is the curing of materials. Most of the systems use at least one wet cement-based material: mortar, grout, surface bonding cement, ready-mix, or shotcrete. To harden properly, these materials must stay moist for about four to seven days. That's because cement actually requires the presence of water to harden. Without it, the final product will be weak and can get surface cracks. So for several of the systems, you need to wait before putting loads on the walls, although you can usually work around them right away. But for others (like the panelized and some mortarless block systems), you can put on a load immediately.

When it affects this curing process, the weather can also be a factor. The trades generally consider light rain to be no problem. But using the wet cement mixes in heavy rain can cause voids and weak spots. Installing before rain arrives is usually fine.

Freezing can rob cement mixes of the necessary water and cause weak spots. Most masonry and concrete crews in cold areas have methods of working in temperatures below freezing. These involve some combination of: adding accelerator chemicals to the mix to cause it to cure before it can freeze; adding antifreeze chemicals; heating the water, sand, or stone put into the mix; and throwing insulating blankets over the final wall. But below 20–25 degrees, we're told, the only option for getting a quality job is to tent in the structure and use heaters. This is usually too expensive for residential construction. In contrast, many of the new CHSs have built-in insulation that allows installation in low temperatures without any special measures.

In hot or dry weather, it's possible for the mix to dry out too fast and end up powdery and weak. But preventing this is easy. The crew covers the wall with plastic sheet or someone wets it down a couple of times a day for 2–3 days.

Crew coordination

A new building system sometimes changes the number of trades necessary and the order of the crew visits. The supervision the builder does might become easier or harder as a result. It's something worth considering so you can pick a system that fits your style of operation, and so you'll know how to run projects smoothly with your CHS.

One concern a lot of frame builders have about CHSs is that the crew installing the concrete might have to make more provisions for later trades. For example, they'd have to preform holes where lines will run through the wall. In some situations, this isn't true with the new CHSs. And in others the crew does need to make some provisions, but there are some newer, easier ways of doing it. This varies so much from system to system that we have to wait for the full descriptions to cover it.

Making connections

One problem with the old concrete systems was that it was difficult to attach components made of other materials. The major connections to the exterior wall are with the flooring and roofing members, interior walls, interior sheetrock, exterior siding, and interior fixtures and trim. Many of these things are easier now with the new CHSs.

One thing to bear in mind from the beginning is that many codes require lumber in contact with concrete to be pressure-treated. This is to prevent the lumber from rotting in case the concrete gets wet for a long period. Several builders we spoke with avoided the extra cost of pressure-treated lumber by simply wrapping their wood with a waterproof membrane wherever it came into contact with concrete. They used #15 felt or heavy plastic sheet for the membrane. Some building departments accepted this.

Many CHSs attach intermediate floor decks with one of three traditional methods: a shelf, a ledger, or pockets. Figures 4-1, 4-2, and 4-3 contain details.

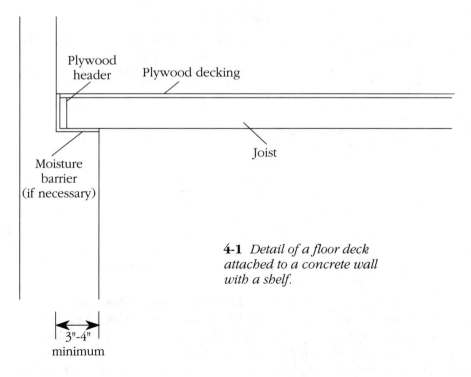

Plywood header

Plywood decking

Joist

Moisture barrier (if necessary)

4-1 *Detail of a floor deck attached to a concrete wall with a shelf.*

3"-4" minimum

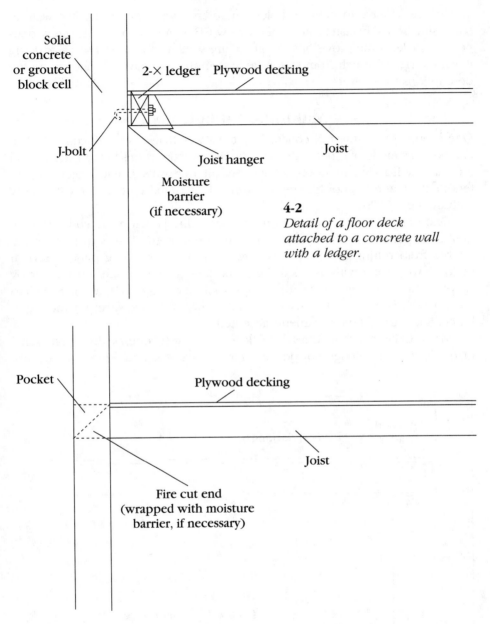

Solid
concrete
or grouted
block cell

2-X ledger   Plywood decking

J-bolt

Joist hanger

Joist

Moisture
barrier
(if necessary)

4-2
*Detail of a floor deck
attached to a concrete wall
with a ledger.*

Pocket

Plywood decking

Joist

Fire cut end
(wrapped with moisture
barrier, if necessary)

4-3 *Detail of a floor deck attached to a concrete wall with pockets.*

Shelfs are possible when the wall steps down in thickness. This is common between the foundation and the first floor. Ledgers require that the block crew install J-bolts at the correct spots, and the pocket method requires that the crew form pockets in advance. But any good block or concrete crew has installed J-bolts and pockets before. If you or the plans just make it clear what method you want to use and where the floor goes, the crew can usually take it from there. Pockets use the least material. But some framing crews don't like them because there's little flexibility to adjust the positions of the joists when putting up the floor deck. Also note that most local codes require a fire cut on the ends of pocketed joists. This is designed to ensure that in a fire they'll fall out instead of pulling on the wall.

Roofs usually go on with hurricane straps or a top plate J-bolted onto the wall. Figures 4-4 and 4-5 show typical details. Hurricane straps anchored directly into the concrete (for a poured wall) or grout (block wall) use little material and make a very strong connection. But some builders claim that they limit how far the roofing crew can adjust the positions of the roofing trusses or rafters. They say that setting a truss as far as an inch to either side of its strap could make the connection too weak to withstand high winds. But some others felt that moving up to 2 inches either way was safe, and that this was plenty of latitude for the roofers.

With most CHSs, windows can go on the same way they do with frame. The wall crew forms a sized opening, then someone sets a window inside it and puts concrete fasteners through the window frame's flanges into the front of the wall. But for most of the CHSs, other options were more popular. Builders' favorites

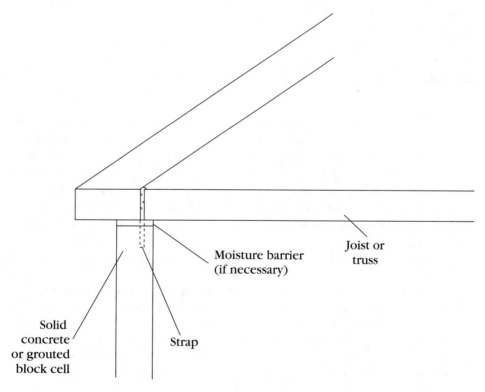

Moisture barrier
(if necessary)

Joist or
truss

Solid
concrete
or grouted
block cell

Strap

4-4 *Detail of a roof attached to a concrete wall with hurricane straps.*

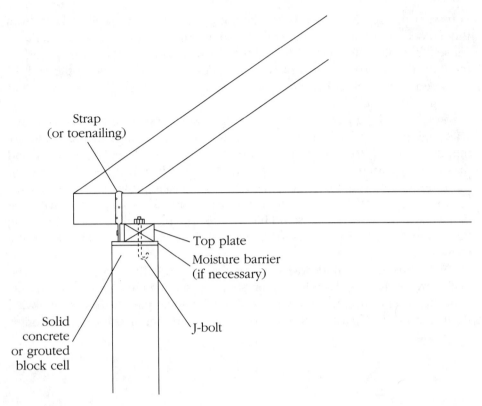

Strap
(or toenailing)

Top plate

Moisture barrier
(if necessary)

Solid
concrete
or grouted
block cell

J-bolt

4-5 *Detail of a roof attached to a concrete wall with a top plate.*

vary a lot from system to system, but one traditional method is used in a lot of cases. That's the *buck*, pictured in Fig. 4-6. A buck is a subframe of 2-x lumber fastened to the inside of the opening. The crew fastens a window with a narrow flange (sometimes called a "masonry window") to the buck, adds sealant around the edges, and covers up to the frame with the siding or trim. Or the crew mounts the buck flush with the outside surface of the wall and attaches the window to the buck just as it would to frame—by nailing or screwing the flange to the lumber.

In most systems builders use the same connection methods for doors they do for windows.

Everything else—interior walls, interior furring or sheetrock, exterior siding, and interior fixtures and trim—often gets attached with concrete fasteners. The favorite fastener tends to depend on the builder, the CHS being used, and the specific item being connected. Table 4-7 lists the fasteners we saw being used most often, along with some information we got from the manufacturers. Figure 4-7 contains photos.

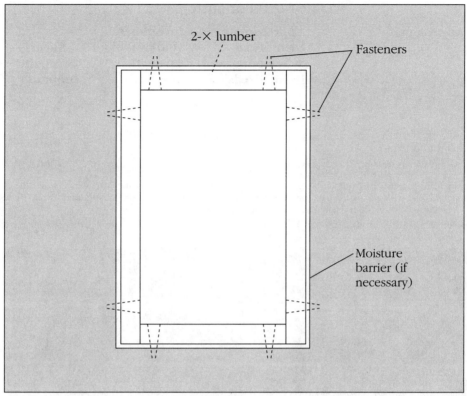

4-6 *Detail of a window buck attached to a concrete wall.*

Table 4-7. Concrete Fasteners

	Approximate Unit Price (when bought in hundreds)	Approximate Pullout Strength (pounds)[1]	Approximate Shear Strength (pounds)*
Drilled Fasteners			
Concrete screws			
1" length	$ .15	300–500	700–800
4" length	.55-.65	1000–1700	1000–2300
Plastic anchors	.05[2]	70–100	500
Metal anchors			
2" length	.30[2]	700–1000	1000–1700
4" length	1.10[2]	2000–6000	2000–7000
Toggle bolts	.40	500–800	550–1700
Driven Fasteners			
Concrete nails	.02	80	600

Table 4-7. Continued

Approximate	Approximate Unit Price (when bought in hundreds)	Approximate Pullout Strength (pounds)[1]	Shear Strength (pounds)*
Powdered-actuated pins			
2½" length	.20	700	700
Pneumatic nails/pins			
light concrete	.03	300	200
heavy concrete	.05	1000	800

1 Strengths depend heavily on the weight of the concrete, the exact design of the fastener, and how the fastener is embedded. In addition, allowable design loads are usually about one-quarter to one-third these test strengths. Consult the manufacturer for precise numbers.
2 This includes the anchor only. A standard wood screw (usually #8 or #10) is additional.

4-7 *Examples of common concrete fasteners: (from left) concrete screw, toggle bolt, cut nail, concrete nail, powder-activated pin and cartridge.*

The "drilled fasteners" require the worker to predrill a hole in the concrete, as is done with a wood screw. But they don't require any special equipment beyond a masonry bit, and most of them can hold a lot of weight. Many masonry screws come with a precisely sized bit included and go directly into the concrete hole. They often go by the nicknames "Tapcon" or "Redhead" after two popular brands. But there are a lot of other brands as well.

The "driven fasteners" go into concrete like a nail without any advance preparation. The most traditional of the concrete nails is the "cut nail," which goes in with a few whacks from an ordinary hammer. Powder-actuated pins sometimes go by the name "Ramset" after one of the popular brands. Each pin comes with a cartridge containing about as much gunpowder as a .22 bullet and is shot into the wall from a special gun. A lot of concrete and carpentry crews have the equipment. Some of the big fastener companies supply masonry nails that you can use with their standard pneumatic equipment, just the way you shoot nails into lumber. For heavier concretes, you might need a high-pressure compressor and special guns, supplied by companies such as Pneutek Corporation of New Hampshire.

Utility line installation

Frame builders generally worry about installing utility lines in concrete walls. They expect all sorts of advance planning, extra crew coordination, and unpleasant cutting. Actually, a lot of the new CHSs have special features that let lines go in pretty much the same way they do with frame. Between the wall construction and the sheetrock, the trades make easy holes in some soft material and set their cable, boxes, and pipe in them. A lot of the other CHSs do use more traditional concrete methods, but without nearly the fuss that most frame builders expect. The specifics vary from system to system, so we leave the details for the full CHS descriptions.

Change flexibility

Few aspects of building with concrete strike more fear in the hearts of frame builders than the prospect of having to change the structure once it's up. They imagine long, unpleasant hours with the sledgehammer, major delays, and high costs every time someone puts a wall on the wrong line or the customer changes his or her mind about where a window goes. But for the new CHSs, the reality is usually far different.

With any building system, structural changes get more difficult and expensive the later you make them. Although framing lumber cuts fairly easily, moving a window even in 2-x-4 walls can be a messy, costly task after the sheathing and sheetrock are up. The real question is how difficult changes become at each point in the building process.

A lot of the new CHSs start with construction of a temporary wall that can be easily rearranged. It's easy to see the whole layout, check it, and make changes. It's only later that wet concrete is added and locks everything into place. And even with the systems that apply wet concrete mixes as the wall goes up, changes are generally much more orderly than frame builders imagine. In the full system descriptions, we cover how builders handled changes at each stage and what the costs were for each CHS.

Bear in mind that the amount of change flexibility you need in a system depends on the way you work. If you build large numbers of standardized homes from the same few designs, you probably don't make many changes. If you're a custom builder who never builds the same thing twice and responds to late buyer requests, you might prefer a system for which late changes are quick and inexpensive.

Manufacturer support

When you're using a new product, help from the manufacturer can save a lot of time and aggravation. The manufacturers of the CHSs provide assistance in one or more ways: written materials, telephone consultation, training, and visits by their representatives to the job site. Exactly what's available varies a lot from manufacturer to manufacturer. In the full system descriptions, we relay what's available for each CHS and how helpful the builders found it to be.

Other considerations

Builders sometimes cited important aspects of a system that didn't fit under any of the other 19 features. When it applies to only one system, we wait until that system's description to explain it. But a couple came up for several systems.

A lot of builders and occupants commented that their finished homes were unusually quiet. When we researched it we found that, yes, concrete walls tend to allow less sound inside than lighter materials do. The ability of a wall to prevent the passage of sound can be measured in the laboratory, and a lot of the CHS manufacturers have it done and give out the results. The number the labs report is the sound transmission class (STC) of the wall. The STC measures the difference between the loudness of sounds on the side of the wall where they originate and their loudness on the opposite side. For example, if a horn blast registers 80 decibels on the side of the wall the horn is on, but only 30 decibels on the other side, the STC of the wall is 50. That is, 80 minus 30.

You have to be careful with STC ratings because the decibel sound scale on which they're based is a modified exponential scale. That means that a wall with an STC of 10 reduces the loudness of a sound by 10 times, a wall of STC 20 reduces it by 100 times, a wall of STC 30 reduces it by 1000 times, and so on.

The STC of an 8-inch block wall with only paint outside and furring strips and sheetrock inside is about 50 (a sound reduction of 100,000 times). The same wall with 8 inches of solid concrete instead of block is about STC 60 (1,000,000 times). By comparison, the STC of a standard 4-inch stud wall sided with clapboard, filled with fiberglass, and covered inside with sheetrock is just below 40. That is, the same outdoor noise sounds about 10 times quieter inside a block wall than it does inside a frame wall, and 100 times quieter inside an 8-inch concrete wall.

Some frame builders expressed concern that there would be a lot of mess to the job site from the CHSs. So we asked about this for every system. The builders almost always said it was no big problem, but we summarize what they told us anyway so you can consider it.

5

Mortared
block systems

Mortared block systems are like a nephew you haven't seen for ten years. When he finally comes to visit, you can see the resemblance to the young boy you once knew. Yet he's changed so much that it's clear he has, bit by bit, turned into a completely different person.

Most of us have been surrounded by conventional frame construction for a decade or more. Our mental image of the concrete block house usually comes from something built in the 1950s. In the meantime, quietly and out of our view, block construction has changed with a series of innovations coming from commercial building and from Europe and Latin America. Builders who suddenly get a look at the latest methods of building with block are amazed at how much it's grown up since they saw it last.

Three of the mortared block systems we found are nonproprietary. The blocks and other components are available from many different manufacturers. What's new is that many of the "other components" that give the final wall its new properties didn't exist a couple of decades ago. But now enterprising builders have combined them to make sharply improved versions of the concrete block home.

The creative mixing and matching of nonproprietary products for block has produced so many variations on the basic block house that we had trouble grouping them into a few logical systems. But after a while it became clear that there was a sharp break between the houses that put their insulation on the inside (which we call the "interior-insulated" system), within the cavities of the block ("cavity-insulated" system), or on the outside ("exterior-insulated" system). Each of these systems has its own strong points, appeals to different market segments, and gets put together somewhat differently.

After the nonproprietary systems, we cover the more radically different, proprietary ones: the Hebel wall system, the Integra Wall System, and Sun Block. For the most part, their manufacturers designed these systems to mesh with the way the traditional trades work. But their materials and some of the construction details are sharply different. As a result, they have valuable new properties that make all of them attractive to several market segments.

The great variation across the different systems makes it difficult to generalize any more. More than any other category, the mortared block systems appeal to a wide range of market segments. Several have come in at or below the cost of frame in many locations and have sold to the value or cost segments of the market. Other block systems are more expensive but have sold to higher-end customers because of their energy efficiency, quality of construction, or striking "look."

On the other hand, the mortared systems also overlap in how they're put together. Some of the basic construction details and field tricks are used over and over again. So, even if you're pretty sure you want to use a certain system, it might also be worth reading about some of the others to get ideas on logistics. In particular, it's smart to read about the first system (the interior-insulated) no matter which one you finally choose. Because it comes first, we cover a lot of the basics of mortared block in its description.

Interior-insulated mortared block

Interior-insulated mortared block is the system that's closest to traditional block. The key differences are new, inexpensive waterproofing techniques, products for putting insulation on the interior cost-effectively, plus a lot of little improvements that have shaved expenses.

Availability isn't a big issue for this system. The new products involved are almost all available nationally, and the traditional components have been available for decades. The trades and skills required are almost all ancient ones still used in every corner of the country.

Cost and exterior finish have been a bigger factor in determining who uses the system. In southern Florida, builders claim they can put up a block house with finishes popular in their area for less than frame. Builders in the rest of the Southeast, in the Southwest, the Northeast, and the Midwest claim they can match frame. Local labor and material costs appear to be competitive in the Plains States as well, but builders there have apparently not warmed up to block because they're not confident they can get an economical exterior finish on it that local buyers will like. The only region where we found no block construction at prices competitive with frame was the Northwest, where concrete is expensive and lumber is cheap.

The basic operation of mortared block is the mortaring and stacking that we all see at commercial job sites. The standard block has nominal dimensions of 8 inches (width) by 8 inches (height) by 16 inches (length). Working with levels and lines and maybe corner poles, masons lay up a series of *courses* (a course is one row) of block. Each course is set on a layer of mortar to bind the blocks and get proper spacing. Figure 5-1 shows a typical residential job site.

The blocks are nearly always laid in a pattern called a *running bond*. In it, each course is offset from the one below it by one-half the 16-inch length of a standard block. It's generally the strongest pattern. A *stack bond*—the blocks stacked with no offset so that the vertical mortar joints line up—is used occasionally for special purposes.

5-1 *Typical block home construction site.*

As they progress, the masons leave openings for doors and windows, using blocks only 8 inches long (called *half block*) as necessary to form square jambs. Window sills might consist of a special precast unit that runs the full length of the opening, as shown in Fig. 5-2. Alternatively, the masons might lay a special sill block. It's usually sloped and solid to cover the cavities of the block below.

5-2 *Precast sill and lintel around a window opening.*

Lintels might be precast units shaped like a U, as shown in Fig. 5-3. The shape leaves a space to put rebar and fill with grout later for strength. We've seen lintels of this type as long as 20 feet go over the door to a double garage.

5-3 *Precast sills (left) and lintels (right).*

The masons can also make their own lintels out of *U-block* (also called bond beam block or lintel block). It's pictured in Fig. 5-4. Its distinctive feature is the cut-down web. The webs are the crosspieces of the block that connect the front and rear slabs of concrete, which are called *face shells*. Because it has these recessed webs, the bond beam block cradles the rebar and lets grout flow horizontally from block to block. To form a lintel, the crew puts up a temporary 2-× frame to support the U-block over the opening, then puts in the rebar and grout.

A final alternative is to lay a masonry-design steel lintel across the opening. The steel lintel has a ridge along the top, giving it a cross section shaped like an upside-down "T." A course of U-block turned upside down can rest on the steel lintel directly, with the ridge sticking up into the block.

If the house has a block foundation, the masons start on the footing. Depending on local structural requirements, they might use a 10-inch or 12-inch-wide block up to the first floor, where they nearly always switch to 8-inch block. When starting on a poured-in-place foundation or slab, they generally use all 8-inch block.

In certain areas of the country, that's the end of the structural work. Where local codes require no reinforcing, you're ready for the roof and finishes. This is

5-4 *A U-block for use in bond beams and cast-in-place lintels.*

true, for example, in some parts of the lower Midwest, where houses aren't often subjected to forces that could break an unreinforced block wall (such as extreme wind, earthquake, or groundwater freezing).

However, in most of the United States, you have to reinforce. This is done by placing steel reinforcing bar and pumping grout into some of the cavities of the block. In high-wind zones, a typical requirement above grade is one vertical rebar every 8 feet o.c., plus one at each corner and one on either side of each opening, plus a horizontal rebar every 8 feet (which is usually at the top of each floor). We call this the 8-x-8 reinforcing grid because it puts one bar every 8 feet vertically and 8 feet horizontally (plus more at stress points). In high seismic zones, the requirement is usually for one vertical rebar every 4 feet and one horizontal every 4 feet—the 4-x-4 reinforcing grid. Below grade the spacing requirements vary a lot, depending on the depth involved and local freezing patterns. Local masons and code officials generally know the rules.

The horizontal courses of block that are filled with rebar and grout are sometimes called *bond beams* because they go around the entire structure, bonding it together. Forming the bond beams starts with covering all the cavities that will not get a vertical rebar with mason's felt, then stacking a course of U-block. Next the crew lowers the vertical rebar into the open vertical cavities and sets the horizontal bar into the bond beam block. Usually, there's also a vertical rebar sticking up into each of the open vertical cavities from the footing or previous bond beam below, as shown in Fig. 5-5. After grouting, the wall will be tied together with continuous reinforcing from the footing to the top plate. After the steel is all in place, the grout crew pumps to fill the open vertical cavities and the bond beam. Then the masons work up to the next bond beam, set the felt and steel, and wait for the next round of grout.

5-5 *Rebar in a lookout connecting foundation to the walls above.*

The simplest reinforcing job is the type done in high-wind zones that have one-story houses set on a slab. Because there's only one bond beam (at the top), the grout can be poured to fill everything at once. This is often referred to as a *monolithic pour*. In high seismic zones, the 4-x-4 grid requires grouting every 4 feet, so there are more pours along the way.

Labor and materials' costs

We spoke with several builders who use both frame and interior-insulated block. Those in the southern two-thirds of Florida claimed that with 1994 materials prices and code requirements, block is break-even or sometimes slightly cheaper than 2-x-4 frame. We also got scattered reports in other parts of the Southeast, Midwest, and Plains States that the two systems are so close in cost that it's hard to predict which will be cheaper for a particular project. Some builders in the Northeast and Southwest said that block comes in about even now with 2-x-6 construction. In the Northwest, the word we got is that the high cost of concrete and low cost of wood make block consistently more expensive.

The big variables that make the picture so different from region to region are the local cost of the materials, local reinforcing requirements, and the exterior finishes involved. In Florida, a standard concrete block can cost as little as $.60 delivered. The reinforcing requirements are moderately high (a grid of 4 feet or 8 feet vertically by 8 feet horizontally). But the most popular exterior finish is stucco, which goes onto block less expensively than onto frame. The net result is that block construction in Florida is about as inexpensive as it is anywhere in the United States and is often less than frame. The story is similar in much of the rest of the Southeast, though not always so strongly in favor of block.

In the Midwest, a block costs more. Ninety cents is a typical figure including delivery. But there's little reinforcing required in some locations. And the builders often compared their block to houses that have a brick veneer on one or more sides. Because a brick veneer is expensive, the block can come out at a lower cost. We got similar reports from the Plains States, except that buyers less often see a block finish as a substitute for brick there.

Block prices also ran around $.90 in the reports we got from the Northeast. There, reinforcing requirements are usually more than in the Midwest. The exterior finish that builders most often compare block to is wooden clapboard, which is usually between stucco and brick in cost. The result is that block construction can cost a bit more in the Northeast, and frame a bit less. The two were about equal in some projects we were quoted in which the frame would have been 2-x-6.

In the Southwest, block is relatively inexpensive. Seventy cents per unit, delivered, was a typical quote. And, as in Florida, stucco is the popular siding. However, reinforcing is on the 4-x-4 grid. The net result is that we were told block now comes in about the same as 2-x-6 construction in many situations. We were quoted costs of about $1000–$1500 higher than 2-x-4 construction in a typical 2000-square-foot (living space) house.

In the Northwest, a standard block can cost $1.20 or more delivered. Reinforcing is on the 4-x-4 grid, and the popular sidings are some form of lapped wood, which is only moderately expensive on frame. The result is that block rarely comes in at a cost as low as either 2-x-4 or 2-x-6 frame.

Table 5-1 contains representative square-foot costs. They're in the middle of the estimates that we received from around the country. Remember that a block covers exactly eight-ninths of a square foot. So when someone quotes you a cost per block, you have to multiply it by 1.13 to get the cost per square foot of wall area.

You can see from Table 5-1 that the range for block costs is a little higher than the range for 2-x-4 frame in Table 4-1. But when you start shifting the costs of materials, reinforcing, and siding to match local rates, practices, and preferences, you find that frequently block comes in lower for a specific job.

You might notice that the materials cost for the unreinforced wall at the top of the table is higher than the cost of a square foot of block (which is 1.13 times $.90, or $1.02.) Part of this difference is the cost of the mortar and part is materials markup by the mason. As with any material, you can save this markup (about 10–20 percent) if you have the knowledge and time to order and receive your own block.

The insulation figures include the new metal-stud-and-bracket systems that are gaining popularity in cold climates. They can actually go to much higher R-values if necessary, as we discuss later.

Among the exterior finishes, the split face is not actually an added finish. It's part of the block itself, formed during the manufacturing process. So the cost listed is the incremental cost of buying this fancier type of block and coating it on-site with a sealer for waterproofing. There are many other so-called architectural block (sometimes called exposed block) finishes used and other methods of waterproofing, covered in the following.

Table 5-1.
Representative Costs of Interior-Insulated Exterior Walls

Item		Cost per Square Foot
Unreinforced wall		
Materials		$1.30
Labor		1.90
Reinforcing		0–1.00
None	$ 0	
8' × 8' grid	.75	
4' × 4' grid	1.00	
Insulation, including furrings strips or studs		$ .20–.70
R-6 wall	$ .20	
R-13 wall	.70	_____
Subtotal, structure and insulation		$3.40–4.90
Exterior siding (materials and labor)		$1.00–9.00
Split face and sealer	$1.00	
Furrings strips and vinyl	1.00	
Stucco and paint	1.10	
Furring, clapboard, and paint	3.20	
Brick veneer	9.00	
Interior sheetrock (materials and labor)		.80
Interior painting		.70
Total, including finishes		$5.90–16.60

We learned that different masons figure their rates in very different ways. Some—generally the larger outfits that do a lot of large commercial jobs—quoted a relatively high base rate per block or per square foot but held it pretty constant. In other words, they wouldn't cut the rate for simple jobs, and they wouldn't raise it for extra corners, extra stories, and so on, unless these things began to get extreme. Other, usually smaller, contractors quoted low rates for a simple rectangular structure but adjusted that up for almost any added irregularity. And some masons used procedures that were somewhere in between these two. This suggests that, for complicated plans, you might want to include a fixed-rate crew among your bidders (because they won't increase charges much for the complexities). For simple plans, you should ask for a bid from at least one crew that prices jobs more on the details.

Most masons would charge more for corners only in extreme situations. On an extremely zigzag structure, one mason told us the total bill might go up 25 per-cent, and another said he might double the charge for the section of wall involved.

Plumbing is another item not always charged for. However, some of the masons with low base rates told us they sometimes double their unit charges on the walls of bathrooms to cover the extra wall penetrations.

The markup for extra stories varied widely in the quotes we got. Some contractors said that they owned the necessary scaffolding, and that their standard rates already included the equipment and extra labor costs of upper-story work. At the other extreme, the highest quote we got was twice the base rate for the second story and three times for the third. A more typical figure was 20 percent extra for higher-level floors.

Curved walls (called *radius walls* by masons) are made by laying half block in a stack bond and angling each column of block a little. The quotes we got for building them were about 25 percent higher per square foot.

Any wall angle can be formed by cutting the block. There are also special blocks for a 45-degree angle that some masons use. Our interviewees were vague about how much they would charge for odd angles, however. They claimed to do few of them and preferred to calculate each job individually. But everyone agreed that the per-block cost went up rapidly when a lot of cutting was involved.

The estimates for reinforcing in Table 5-1 show that the cost jumps at first, then rises more slowly as you add more. So, for example, doubling the reinforcing doesn't usually double the expense. When most masons quote their standard rates, they're including the cost of the grout and the labor to do the amount of reinforcing that's standard in their area. Their quotes rarely, however, include the rebar. By tradition the general contractor provides that, so don't forget that the mason might be assuming you'll buy it. Number 5 rebar runs about $.15 per square foot on a wall with a 4-x-4 reinforcing grid.

Learning costs

A builder who learned frame and block construction at the same time told us that he learned them equally quickly. He felt comfortable with each one after doing four to five houses with it. Frame builders who started using block later agreed that the two systems are of about the same complexity. But they felt comfortable with block after only a couple of houses. They claimed that a lot of the principles are the same as with frame, so they had a head start.

Every builder claims to have spent a sizable amount of personal time talking with subs and vendors and doing reading to plan out their first couple of block houses. But this time blended in with their other activities. As one put it, "I guess it might have slowed me down on some other things, but it might have revved me up, too. It was fun. It was sort of my hobby for six months."

Several said that they overbuilt the first house. One builder estimated that excess materials and labor cost him around $2000 on his first block house, out of a total cost of construction of about $100,000. Some extra expenses we heard cited included using top plates and window bucks that were larger than necessary, and having the masons install extensive flashing in the block because of fears of water penetration. In addition, some crews work slower at first, especially in installing some new interior insulation products that are unfamiliar.

The electrician was the subcontractor that was most often hesitant to work on a block house for the first time. But, as we discuss in the following, in the interior-insulated system the electrician has to deviate very little from normal practices. Some of the electricians who were asked to make bids refused to do so, and some came back with estimates that included increased charges of a couple hundred dollars or so. But usually the builder was able to talk them out of increased charges by demonstrating the ease of running cable. We were told that the electrician's work generally turned out to be "easier than expected."

Buyer reaction

The way buyers reacted to the information that a house was built of block was very sensitive to the way the house looked. When simply hearing about the material, some buyers in the Northeast and Midwest had a negative impression. But when they saw the final product with one of the modern finishes and conventional sheetrock inside, that changed. As one house hunter in Indiana put it, "I didn't know you meant *that* kind of block." The lesson one builder gave us was, "Always show them something. Just describing doesn't convince anyone."

The story was different in the Southeast and Southwest, where there's a greater tradition of block homes. Frequently, buyers there told us they had a favorable impression of block as a more solid material. They considered it a plus when comparing homes. There were also buyers who were never aware of what the structural material was. With stucco outside and sheetrock inside, the difference was not apparent, and they didn't ask.

Hearing buyers' reactions drove home a couple of important points that surfaced in other places. Being successful with block in the value segment depends a lot on finding an exterior finish that fits local tastes and can be put on block without driving costs above those of frame. When builders found an appealing, economical exterior, buyers tended not to be concerned about the structural material.

Exterior and interior finishes

Every interior-insulated mortared block house we encountered used sheetrock as its interior surface. The insulation and studs (or furring strips) on the inside have to be covered, and sheetrock is the most popular and one of the least expensive ways to do it.

In the Southeast and Southwest, 90 percent of the block houses we heard about had stucco outside. It's popular in those areas and it goes onto block inexpensively. In both areas we received quotes as low as $.50 per square foot, labor and materials, plus $.40 to paint it, for a PC stucco. Some builders paid the extra for PM or PB stucco because of the reduced cracking and final color coat. Some paid the extra for an elastomeric paint because it covers cracks.

In the Northeast and Midwest, stucco is popular in spots, but not as consistently as in the two southern corners. In these regions a more popular choice was some type of architectural block. Architectural block is specially manufactured to get a particular appearance on one or both sides. There are dozens of types, each with its own look. Some of the more common ones are pictured in the color plate section at the center of the book. Prices for the different types (in the middle of quotes we got from around the country) are in Table 5-2.

Table 5-2. Representative Costs of Architectural Blocks

Style	Cost per Block[1]	Cost per Square Foot
Smooth (standard, nonarchitectural block)	$ .90	$1.01
Split-face	1.30	1.46
Scored	.95	1.07
Half-high (also called "four-high")	.65	1.46
Ribbed	1.15	1.29
Split score	1.35	1.52
Split half-high	.80	.90
Slump half-high	.85	.96
Ground-face	4.00	4.50
Glazed	5.00	5.63
Colored	+.10	+.11

[1]All figures include a typical shipping charge, but not a mason's markup.

Split-face block is "broken" in the plant so that its exterior face is rough, like a cut stone. Scored block has one or more vertical lines to make it look like a set of smaller units instead of one that's 16 inches long. Blocks with a single score look like two 8-inch square units side by side. Blocks with more scores look like narrower units stacked on end.

Ribbed block has so many scores that it has a series of thin, vertical edges (or "ribs") that appear to run up the length of the wall. Half-high is only 4 inches high, and therefore looks something like a long brick. Any of these shapes—scored, ribbed, or half-high—can also be split to give their surfaces a rough texture.

Slump block is made with a thin concrete so that it sags a little before it's cured. It has a look similar to handmade adobe. Ground-face block is polished after hardening to give it a granite or marble appearance. And glazed block is surfaced with a glaze much like a ceramic tile. The result is a smooth, shiny surface in almost any color.

It's also possible to put color in the concrete before making any of these blocks (although there's not much point in doing so for a glazed block). In these cases, the masons are usually instructed to add a pigment to the mortar as well to match or contrast with the block's color.

Color is popular with exposed block and greatly expands the range of appearances that are possible. It adds about a dime to the cost of a unit. But beware. We consistently heard that you can get great colored block if you know what you're doing, and problems if you don't. The most conscientious block producers do a good job without your saying anything. But concrete coloring is sensitive to correct measuring and thorough mixing in the production process— more so than almost any of the other mixing work that block plants do. Some inexperienced plants have been said to deliver block with uneven or incorrect coloring. Some careful masons get agreement up front from the plant that it will reimburse them for any incorrectly colored units, and that if there are too many of these they'll rerun the whole order. They also make sure the plant produces all the block from one batch of concrete to eliminate the possibility that they'll get two runs of block with different concentrations of coloring in them. They might even overorder to be safe. If you want some variation in color, the safest bet is to go with a plant that has experience at this and will make good on any splotchy units that it delivers.

The builders in the Northeast and Midwest almost always opted for a split-face block, a half-high, or a combination of the two (the split half-high). The best that we could determine, the appeal of split-face block is that it reminds people of the stone buildings of centuries past, many of which are still standing in these parts of the country. They have a traditional American charm and an appearance of durability. Half-high has strong appeal in regions accustomed to brick.

Brick is also the siding of choice in much of the Plains States—roughly from Nebraska to Texas, to be more precise. However, block has seldom been used there to date. So we have no good read on how half-high or any other type of architectural block would go over in that area. Half-high slump block is used in parts of the Southwest and West Coast to get an adobe look. However, it's almost always used in a cavity-insulated system, so we discuss it in the next section.

Builders' attraction to using architectural block is clear. With it, the siding comes at a low incremental cost and without an extra crew. Consider split-face block. The block itself costs an extra $.45 or so per square foot, $.56 if it's colored. As we'll see later, waterproofing is necessary but can be purchased for as little as $.20 per square foot. To get it to look really good, the masons have to take extra care in laying the block, tooling the mortar joints, and keeping the wall clean. A high-priced mason we interviewed said that he charges an extra $1.00 per square foot for this, but more typically we got quotes that were a fraction of that. The result is that for about a dollar per square foot, you have a siding that's nearly zero-maintenance and attractive to many buyers.

If you use half-high, the cost is greater. The masons have to lay twice as many units. And the materials cost will be around $.45 higher per square foot because half-high is more than half the cost of standard-size block. Quotes from masons varied widely, from about $.75 to $3.00 extra per square foot for the materials and labor. But even with the higher rates, the cost is usually half that of a brick veneer. Including the costs of waterproofing and careful finish work, half-high can be produced for $2.00–$4.00 per square foot over the cost of a standard block structure.

We suspect there's room for a lot more creativity in the use of architectural block. It makes sense that some types other than split and half-high might sell in the Northeast and Midwest, and that some types of architectural block could be attractive to customers in the Plains States and the regions dominated by stucco. As one builder put it, architectural block, "has the advantage of being different. A lot of people are getting tired of the same old things we've been giving them for the last ten years." We found it was very popular with architects. They put together some striking combinations of contrasting colors and textures. Even in parts of the country where we found little use of block overall, we stumbled across a few architecturally designed houses of exposed block.

Interior-insulated block can also take virtually any of the other sidings traditionally attached to wood: clapboard, shingle, vinyl, aluminum, hardboard. These are typically attached by first putting furring strips on the block to form a backing for nailing or screwing. The cost of the furring materials and labor averaged about $.20 per square foot. While not great, this is an extra expense to using block when you need one of these sidings. We didn't find them used often. Of course, block can also have a brick veneer. This results in a double-wythe wall. The cost of brick is about the same over block as over frame.

Design flexibility

Block can actually do just about anything frame can, if you throw in some 2-x lumber here and there. Block is widely used in basements and above grade up to several stories. The only thing that frequently changes with extra stories is the reinforcing requirements. They might increase the farther you go up or go down. You have to check the local code on this.

Odd angles and curved walls are done as discussed under costs. Several builders told us they found it less expensive simply to do curves or sections with odd angles (like bays) out of frame. This will again vary with how much your masonry crew charges. It's also important to remember that block generally needs a foundation below it. If you have a bay that's cantilevered, for example, you probably need to do it in frame.

For irregular openings, such as arched doorways and diamond windows, almost everyone we talked with built a standard square opening out of block and a lumber subframe to fit into the opening. The subframe made the odd shape and attached to the block like a buck.

We heard one major design consideration over and over again: Make sure the house is designed on an 8-inch module. That means that the lengths and heights of the walls and the positions and sizes of the openings should all be some multiple of 8 inches so the walls can be built out of whole and half blocks without cutting. Lots of cutting slows work and increases costs.

The builders told us that design on the 8-inch module isn't hard to do if you're prepared for it. You tell the buyer and/or architect as soon as possible that this is how dimensions should be determined and get the plans done accordingly. It puts a constraint on design, but some architects we talked with liked it. They felt it put order on the work and lent consistency to the final appearance. One said, "I put a grid on the screen with my computer design pro-

gram and set everything on the lines. It looks great, and the clients love it."
Where deviations from the module are important, you can have the masons do
some block cutting or use frame.

R-value and energy efficiency

The insulation in this system is added to the interior of the wall after it's erected.
An 8-inch-wide block itself has an R-value of 2–3, depending mostly on the
weight of the concrete used. In some very warm areas, 1-x-2 furring strips are
attached inside with powder-actuated fasteners or masonry nails, and insulation
is placed between them. This produces a nominal wall R-value of 5–8. Breaks in
the insulation that would lower the R-value are the furring strips and any plumbing, wiring, and boxes run between the strips.

When they needed a higher R-value, builders almost always used metal stud
systems in place of furring strips. You could use wood, but we were told the
stud systems are cheaper.

There are two basic types of stud systems. The first uses specially manufactured studs that hold extruded EPS foam board insulation against the block with
a screwed fastener. The studs attach with metal brackets so that a minimum of
metal touches the block. With this arrangement, the insulation breaks formed by
the studs and brackets are kept small. Figure 5-6 contains a diagram of a typical
system.

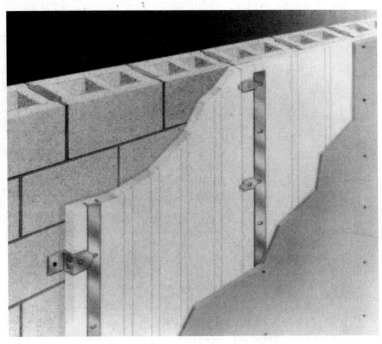

5-6 *Diagram of metal stud insulation system.* W.R. Grace

Special stud systems cost about $.50 per square foot for materials (assuming 2-inch foam) and $.20 for labor. Manufacturers are listed at the back of the book under Metal Stud Insulation Systems.

The second type of stud system that builders told us about consists of standard interior steel studding. One rail is fastened to the floor and one to the ceiling joists. Steel studs are fixed to the rails 16 inches or 24 inches on center, and insulation (usually fiberglass) is put between them.

According to data from the American Society of Heating, Refrigeration, and Air Conditioning Engineers (ASHRAE), the effective R-value of a steel stud wall is about half the R-value of the insulation itself. For example, a wall of four-inch studs placed 16 inches o.c. with R-11 insulation rates about R-5.5 by the parallel path method. For that reason, builders told us they set the rails 1–2 inches away from the block and put foam board in between. An inch of foam and the 4 inches of fiberglass between the studs should yield a wall of about R-14 (isothermal planes method), including the block and sheetrock.

Using standard steel studding has the advantage of using some lower-cost materials: ordinary studs and fiberglass. However, there is also the added cost and labor of adding foam board. Standard studding results in a thicker inner wall—5–6 inches versus the 2–4 inches of the special stud systems, on top of the 8 inches of the block. With either system, there are occasional breaks in the insulation from wiring and plumbing.

Standard block has a weight of 35–40 pounds and a thermal mass of about 8 Btu per square foot per degree Fahrenheit. The wall's mass is approximately double this where it's reinforced. This is enough to entitle interior-insulated block to having its R-value counted higher under almost all codes that count thermal mass.

Thermal mass

This mass is all on the outside of the insulation envelope. It therefore contributes little to the sort of energy storage used in passive solar or load-leveling houses. The interior thermal mass of the wall is the sheetrock, just as it is for frame.

Water resistance

When the wall is sided with stucco, the stucco and paint are usually the primary moisture barrier. In cold climates, we frequently heard builders tell us they used an elastomeric paint or PM or PB stucco for extra protection.

We spoke with the occupants of some stuccoed homes that were a decade old, but still used modern-type stuccoes and paint. Some were in wet southern climates. They all told us they noticed no water through the walls and did not believe their homes were particularly humid.

Houses with architectural block used either a spray-on sealer or a water-repellent material added to the concrete. Painting contractors put sealers on.

They're inexpensive (as low as $.20 per square foot for labor and materials), and they need to be reapplied every 6 years or so. Good contractors can usually make a recommendation.

Water-repellent materials are put into the concrete at the block plant, and into the mortar by the masons. They increase the masonry bill about $.30 per square foot. Builders who've examined the treated block say it's impressive. Water poured onto it simply beads up. In addition, no maintenance is necessary; the repellent is a permanent part of the block.

Builders warned us to make sure the masons also put a water-repellent material in the mortar. Crews that have used it before can usually be trusted to include it in each mix, but there are reports of others who occasionally skipped it here and there. So spec it clearly, and with a crew you don't know it might be wise to check up. Many masons are familiar with it now.

Some builders using an exposed block exterior had their masons put flashing and weep holes in the bottom course of block above grade and above the lintels. This is a common practice in commercial construction to drain out water that happens to get through the exterior face shell. It might add one or two hundred dollars per house. On the other hand, some other builders felt this was overkill.

The occupants of houses with exposed block and a sealant or water repellent also reported no water problems in their walls. This held true in northern climates with frequent freeze-thaws, as well as warm climates. However, most of the houses are still fairly new. Still, the early signs are good.

Maintenance required

Our builders and homeowners confirmed that stucco and exposed block siding require less frequent painting or sealing than wood sidings. Paint or sealer needs replacement every five to seven years.

Supposedly, block containing a water-repellent agent won't need any further maintenance for decades. We can't confirm this because we didn't find any houses using it that were that old. But it makes some sense because the agent is an integral part of the block and mortar, not a surface coating.

Some builders in warm climates who use traditional PC stucco told us that hairline cracks can open up in the half year or so after the house is built, as the foundation settles. This was a more frequent occurrence in areas with soft soil. Some of the builders plan on it. They routinely send a crew around to all houses six months after construction to fill in cracks with an elastomeric grout and paint over it.

In the northern climates, fewer of the houses we heard about used stucco. But when they did, the builders and trades favored a PM or PB, or an elastomeric paint. We were told that these minimize cracking over freeze-thaw cycles and make follow-on maintenance rare. No one reported any maintenance needs on the structure itself.

Disaster resistance

Plain 8-inch block walls usually get a two-hour fire rating, after which time the cool side of the wall passes the temperature limit allowed. The test is almost never stopped because of a structural failure.

As noted in chapter 4, block walls are generally conceded to endure fire better than frame. In some areas insurance rates are lower as a result. Also as described in chapter 4, block homes appear to have survived Hurricane Andrew better than frame. Most of these were probably reinforced on an 8-foot-x-8-foot grid. If you want to be extra safe, you can increase the frequency of the rebar.

As for earthquake, a 4-foot-x-4-foot reinforcing grid is supposed to be sufficient for zone 4. Unfortunately, we have no actual results to compare the survival of 4-x-4 with frame. These should be available when detailed studies of the Northridge quake of 1993 are done.

Product availability

All the builders and masons said that almost everything that goes into an interior-insulated block house is easy to get. Most architectural and colored blocks have to be advance ordered, but a couple of weeks should be enough time.

Block plants vary some in the range of product they produce. You can count on just about any of them providing color and making split-face, half-high, and scores. Most will also sell ribbed. Slump block, ground-face, and glazed require special procedures and equipment, so they're rarer. Slump is made in the Southwest but not much in other regions. Ground-face and glazed are made here and there by larger plants throughout the country. You can always get any block if you're willing to pay shipping costs. But the extra money gets to be too high for most projects when the nearest plant is more than a couple of hundred miles away.

Some of the specially designed metal studs for interior insulation have to be shipped, but the cost is usually minor. You just have to order in advance.

Water-repellent agents are also widely available from block plants. The ones that don't carry it can order it.

Labor availability

Everyone agreed that there are plenty of masons who can build the walls to an interior-insulated house. But they often had to shop to find ones that were well-suited to the task.

In regions where block houses are built regularly (primarily Florida and Arizona), builders just find crews that do a lot of houses. In other regions, there might be a few choices. Often there are masons who do brick veneer on homes and block on commercial buildings. They usually have the right mix of skills. In areas where basements are commonly block, some builders used their basement crews. As one builder put it, "We just asked them to continue up to the roof line. They like it because it was more work for them, and they did a good job for us."

We heard several pluses and minuses to using purely commercial crews. The quality of their work, by all accounts, is high. They also tend to be set up to do just about anything you might want: odd walls and openings, unusual reinforcing patterns, and so on. On the other hand, for simple designs their rates tend to be relatively high. We also heard that they tend to be somewhat less flexible in their work practices. But, according to one builder, "These guys all work independently, too. You can hire some in their spare time."

Most areas also have plenty of stucco crews if you need one. Builders in cold weather areas generally took more care in making sure they had a crew that knew its materials and installation practices well. The metal stud systems are usually not difficult to use. In some cases, carpenters put them up. One builder that used one read the instructions and got a half days' training from the manufacturer. He found that he and an unskilled helper could put it up easily.

Building department approval

Block walls were always familiar to the local building departments. We found no cases of departments requiring product documentation or an engineer's stamp for interior-insulated block houses or the interior metal stud systems.

In most localities where reinforcing is required, an inspection is done at each bond beam to make sure the appropriate rebar is in place before the grout pour. If horizontal reinforcing is required every 8 feet, this usually means that an inspection takes place at the top of the first floor and again at the top of the second (if there is a second). To make it easy for the inspector to see that the vertical rebar is in place, the masonry crew in some areas cuts a set of holes in the blocks. These holes are called *lookouts* or *cleanouts*. One is pictured in Fig. 5-5. They get patched over before the grout pour.

In the case of a 4-x-4 reinforcing grid, the inspectors come out every 4 feet of wall height after the steel is placed. We thought this might be inconvenient. But builders in areas where it applies told us that the local building departments are experienced enough at this arrangement that they generally know how to time their rounds and keep the delays limited.

Required calendar time

In Florida, we saw a crew of six masons and seven laborers put up two 1800-square-foot (living space) ranch homes from slab to top plate in one day. The foreman told us that by adding a couple of extra people they could knock out even a 2400-square-foot ranch in half a day. This included placement of the rebar, which goes quickly. The developer they worked for had a separate grout crew that followed a day or two later to do the pours. Grouting takes about half a day per house.

This crew was so fast partly because they built a few more or less standard plans over and over. The location of a couple windows and pipes was about all that changed. Other masons told us that laying the walls of a custom house would take about four days with a crew of five. In a very large two-story house, they might count on one week for the first floor and one for the second.

Most thought block walls would require about the same calendar time as frame with the same size crew and same structure. If the block had to be reinforced, there would be perhaps one additional day of waiting for the mortar to set and then a half day of grout work. If the block house had an architectural exterior finish, there was no siding work. This saved a day or so compared with frame.

We were told that roof trusses could go up a couple of days after the block had been laid (the day after grouting if the walls are reinforced). Some builders had tricks to start framing even before that. Some had the carpenters work on other things that don't put weight on the structure, like interior walls. The large production builders simply cycled their crews so that the carpenters trailed the masons by two days.

The most common sources of delay were rain and cold. Some crews said they could lay block in moderate rain, but usually didn't because the finished wall tended to look bad afterward. As noted in chapter 4, some crews can work in weather as low as 20 degrees.

Crew coordination

With interior-insulated block, there are few special considerations in coordinating crews. Wiring and small piping goes through the furring strips or interior studs. The electrical and plumbing crews place this after the block is up. The masonry crews don't even need to know where the runs will go.

Plans sometimes call for pipes or vents to go inside the block. In this case, they usually are sticking out of the foundation when the masonry crew starts. The masons thread the block over or around the lines, as shown in Fig. 5-7. For a vent, which goes all the way to the roof, the usual method is to use a pipe that comes in lengths of 2–4 feet. Masons can thread over a short length, then add on to the vent when they come to the top.

Furred walls occasionally require holes in the block for deep electrical boxes. Experienced masonry crews read these in the plans and make the necessary cuts as they go. Or some electricians will knock out holes as necessary.

If the exterior will be stuccoed, the masons need to be told to leave a stucco-ready surface. To do this they strike the mortar flush with the block. The masons also have to leave J-bolts for floor deck ledgers and bolts or straps for top plates. These things should be on the plans.

Making connections

The popular methods of connecting floor decks and roofs are exactly as described in chapter 4. Door frames are almost always attached by way of a buck. The carpenters install the buck, then shim and screw in the frame just as they would on 2-x-4 construction. Sealant goes at the joints. If stucco is used, it goes

5-7 *Threading block around plumbing lines.*

over the buck and the edge of the sealant. The buck is usually attached with one of the standard screwed or driven fasteners. Almost any one will do. Builders who wanted extra strength or security used J-bolts.

With windows there are more options. In many cases, you can use a window designed for frame and nail its flanges to a buck that's flush with the exterior, or use concrete fasteners to screw directly to the block. If you're using an architectural block, this is less feasible because there's no siding to cover the flange. Also beware that a lot of windows designed for wood have dimensions that aren't in increments of 8 inches, so you might have to modify a buck to adjust.

More often you use one of various types of masonry windows. These are designed to fit wholly within an opening sized for the 8-inch module. The first type has narrow flanges to nail straight into a buck. If the exterior is architectural block, the buck is set back from the exterior surface and trim to cover the joint attaches to the buck.

For another popular masonry window, the masons lay a special block along the jambs that has a groove up the side. The grooves line up from block to block along the entire height of the jamb. Before the masons put on the lintel, they slide the flanges of a special window down the groove.

The last popular masonry window has no flange at all. It's sized to slide directly inside an opening. Some of these windows have metal clips sticking out from the frame. When the masons are a couple of courses past the sill level, they set the windows in the half-formed opening and stick the clips into the mortar joints, then continue up with their block. Others simply have screw holes pre-formed at intervals around the frame. The masons leave the opening, and the carpenters set and shim the window and attach it with concrete fasteners.

Inside, extra furring strips or studs are run around the openings to provide a nailer for trim and sheetrock. A correctly sized buck can act as a nailer right along the window frame.

Interior walls usually get attached simply by shooting the end stud into the block directly with masonry nails or powder-actuated pins. A few builders used J-bolts. Furring strips also went up with nails or pins. The specialty stud insulation systems come with recommendations on fasteners. Interior fixtures and trim attach to the furring strips or studs the same way they would in a frame house.

Utilities installation

As discussed under "Crew coordination," most utilities go in the same way they do with frame. The only difference is that they go through furring strips or metal studs instead of wooden studs. The only lines that frequently run inside the block are larger ones, like plumbing and vent stacks, as already described.

Change flexibility

Most builders' method of handling changes in the block structure is to plan ahead so that they aren't necessary. They tell clients to have their minds made up about where windows and doors go in advance, then they stick by their guns. One said, "After going through the pain of moving a window once, I had the backbone to make my customers do their planning and not let them make changes whenever they felt like it. And it got me to do my planning, too."

When they did make a change, the builders agreed that it was usually more difficult and expensive than with frame. If it came in the first couple of hours after the block was laid, the masons could still knock out and move block around. After the mortar sets, workers made or moved openings by cutting along the mortar joints with a power saw fitted with a diamond blade, then knocking out the loosened blocks. They would even cut out the half blocks on the side of an opening that needed to be built out so that they could "sawtooth" the new blocks into the old. Inserting or moving a lintel over an opening could be a little tricky because the blocks above have to go unsupported for a while. So for large openings, some type of brace is necessary.

We got quotes for moving a window opening that ranged from $200 to $500. If the grout had already been poured, this went up to as much as $1500, including the cost of cutting through rebar and hardened grout, paying an engineer for a new reinforcing design, then getting into the wall at other points to fill in with new reinforcing to fit the new design.

Manufacturer support

The products for this system are so common that the builders and trades don't ask for help often. When they do, it's usually at the local block plant. Most plants have people with a lot of experience who are willing to answer questions.

A builder who used one of the special metal stud systems for the first time claimed he could get a salesperson to visit the job site for the first half day or so of installation. He found this helpful. But he also said that the product was pretty easy to figure out with just the written instructions.

Other considerations

Everyone said that mess to the site is not a big problem. Mortar droppings fall, but inside they can be swept off the slab. Outside, conscientious masonry crews chop them up and work them into the soil at the end of the day. Block pieces can be dumped with the rest of the rubbish. Some builders used them as fill in low spots. This is usually acceptable because they're stable (unlike wood, which rots) and nontoxic.

A few people pointed out to us that it seems that less noise from outside penetrates block houses, compared with frame. This is consistent with the STC ratings listed in chapter 4.

Cavity-insulated mortared block

Putting the insulation inside the cavities of the block eliminates the need for sheetrock and furring or studs inside. It also exposes the block to the inside air, providing interior thermal mass for solar and load-leveling applications and general indoor comfort.

We were surprised by the market segments that cavity-insulated houses were being sold to. We had expected them to be popular in the cost segment. By simply sealing or painting the block on both sides, you can get an inexpensive structure.

But we found cavity-insulated mortared block used at least as often in homes for the quality segment. The reason was aesthetics. A block exposed on both sides can have a distinctive and interesting look. You can't get it economically with frame, and it sets the house off from anything else in the neighborhood.

Cavity-insulated homes for cost and for quality buyers are clearly two different beasts. Many construction details and the costs differ widely. Most of the mechanics of building a cavity-insulated house are the same as an interior-insulated house. The two biggest differences are the way the insulation is installed and how utility lines are run.

Historically, like all block, cavity-insulated homes were more common in the southern half of the country. This still appears to be true for units intended for the cost segment. But fancy custom versions are also popping up occasionally in wealthy neighborhoods of northern states. This has become possible because of new, high R-value insulation products.

Labor and materials' costs

Table 5-3 gives cost estimates that were in the middle of the quotes we got. They vary a lot, depending on how fancy you want to get. Simply laying and reinforcing the block costs the same as with interior-insulated construction.

Table 5-3.
Representative Costs of Cavity-Insulated Exterior Walls

Item		Cost per Square Foot
Unreinforced wall		
Materials		$1.30
Labor		1.90
Reinforcing		0–1.00
None	$ 0	
8' × 8' grid	.75	
4' × 4' grid	1.00	
Cavity insulation		.50–2.00
R-6 wall	$ .50	
R-14 wall	2.00	
Subtotal, structure and insulation		$3.70–6.20
Finishes (materials and labor)		$1.00–4.20
Simple: Sealer outside	$ .20	
Paint inside	.70	
Fancy: Colored half-high	3.00	
Sealer outside	.20	
Premium mortar work	1.00	
Total, including finishes		$4.70–10.40

If you were building the cheapest possible wall, you would finish outside with a sealant or paint for as little as $.20 per square foot and inside with paint for another $.70 or so. The insulation adds about $.50 per square foot when a low R-value is sufficient, and maybe $2.00 when you need a high one. The total is thus $4.70–$6.90, depending on reinforcing and insulation.

For the quality buyers, the block is almost always exposed on both sides. It's frequently a half-high, colored, and sometimes split on one or both sides. For example, a colored half-high might add about $3.00 to the cost of the block. The sealer is still necessary, and you'd need careful mortar work to get a really attractive wall. The masons might charge another $.50 per square foot per side for this—a total of $1.00 in this system. The total would be $8.60–$10.40, depending on reinforcing and insulation.

The builders we spoke with found it hard to beat cavity-insulated construction for the cost segment when low levels of insulation and reinforcing were acceptable. Some quoted savings of around $500 for a 1000-square-foot home, when comparing to 2-x-4 construction and vinyl siding. Others found it about even. Where reinforcing was necessary, the two systems were about the same price in some locations. We found no cavity-insulated construction used in affordable housing projects in cold climates.

In quality segment projects, there was no easy way to make a cost comparison. The buyers wanted block because of its appearance, and there was no practical way to get that appearance with frame. Everyone agreed that the cost, all things considered, was "significantly above" the cost of frame with stucco or wood siding. We also asked for a comparison to a full brick veneer house. Most agreed the block should be cheaper, even with 4-x-4 reinforcing and insulation for a cold climate.

Learning costs

Most of the extra up-front costs we heard described were the same as they were for interior-insulated block. But one was somewhat more severe. Because the interior is generally not furred and sheetrocked, there could be more utility lines run through the block itself. Builders of low-cost houses usually just changed their designs so all possible lines ran through interior walls. But builders of high-end homes didn't always have this option. Builders inexperienced with building block houses with a lot of embedded lines usually spent more time coordinating the trades. No one could really put a price on this for us. The general recommendation was to be aware of it and plan ahead carefully on the first house or two. Another option was to go with a very experienced, high-quality masonry crew and make sure the plans called out every line clearly. This can be expensive, but it was usually the route taken on the high-end homes. The expensive crew was necessary for many other details anyway.

Buyer reaction

In private projects, the reaction depended on who will live in the units. Some residents are simply happy to have a home of their own, and the material and aesthetics are secondary. From what we were told, when the residents are in a higher income bracket, they tend to lean to a more conventional structure. The block surface on both sides—even when smoothed by a thick coat of paint—is not attractive to them. However, they weigh that against the savings.

In affordable housing projects that are government regulated, this system is not always allowed. Most regulations are designed to produce a house that looks more or less conventional.

In the quality segment projects, a positive reaction was a foregone conclusion. The buyers or architects requested block because they already liked it. When builders suggest it, some buyers bite and some don't. However, buyers in the regions that haven't used much block for houses were prone to asking questions about cost and water penetration.

Exterior and interior finishes

In affordable housing projects, the common finishes are a thick coat of paint inside and out, or sometimes stucco outside for an extra $.50 or so per square foot. In custom homes the preferred finish is whatever the buyer wants. This usually turns out to be a colored split-face, half-high or split half-high. On the West Coast, slump block is sometimes popular. After laying the slump block, the mason makes a slurry of mostly cement and applies it to both sides with a heavy cloth. By using a circular motion he can get the appearance of adobe. However, unlike adobe, the wall can be insulated and reinforced. Quotes on this type of wall, complete with insulation, reinforcing, and finish on both sides, ran upwards of $10.00 per square foot.

The inside finish in custom homes was occasionally different from the outside. For example, some houses used a colored block that was split on one side only. And sometimes walls in the kitchen or formal rooms were furred out and covered with sheet rock.

Design flexibility

The costs of unusual structural features are about the same as for interior-insulated block. There was rarely much use of lumber to form odd angles or openings, however. In homes directed at the cost segment, such features added more cost than was acceptable to buyers and occupants. In the quality segment, frame would clash with the exposed block finish of the rest of the house.

R-value and energy efficiency

For low R-values there are three common types of cavity insulation: poured-in-place, foam, and standard inserts. Poured-in-place insulation consists of a light, flaky or pebble-like material that a separate crew pours into the wall after it's up. In an unreinforced wall, it can be poured into the top before the top plate goes on. In a reinforced wall, it can be poured through temporary holes cut in the block, or into the top just before each bond beam is grouted. Pouring before each bond beam takes close coordination of grout and insulation crews. But in an expensive house, it might be preferable to having to hide access holes.

The two major varieties of poured-in-place insulation are vermiculite and perlite. Perlite is available around the country from hundreds of small, independent companies. Vermiculite is distributed nationwide by W.R. Grace Company. Crews that provide and install these insulations are listed in most local construction services directories and Yellow Pages. Grace is listed under "Pour-in-place insulation manufacturers" in the back of the book.

Foams are also installed after the wall is up, through the same types of openings. Most consist of two liquid chemicals that react when they come into contact. The installers squirt both simultaneously into the block, and the mixture foams up to fill the cavities. Foam crews are also listed in local directories. There are a lot of manufacturers. The larger ones we're familiar with are listed in the back of the book under "Foam-in-place insulation manufacturers."

Inserts are pieces of molded EPS bead that are wedged into the cavities. Usually the block plant puts them in and ships the blocks (which they call standard pre-insulated block) to the site. The masons stack them up like ordinary blocks, and when they're done the walls are already insulated. To find block plants that supply them, you can call local plants directly, or contact some of the half-dozen manufacturers of the inserts. They're listed in the back of the book under "Concrete block pre-insulation manufacturers." Most sell regionally, so you might want to call the ones nearest you.

R-value claims on these cavity insulations differ widely. Our own quick isothermal planes calculation suggests that no amount of insulation, no matter how high an R-value it has, is likely to produce a wall built of standard block with an R of over 8. The webs of the block let through a certain amount of heat, and this can't be reduced by putting more insulation in the cavities. The more conservative companies claim that walls with their products have an R-value of 4–6.

Builders' opinions vary a lot about which is the best type of insulation to use. The pour-in-place and foams fill the cells completely, but they can't cover grout-filled cavities. Inserts can. The first two also don't affect the masons' work, but can leak out of holes. The old foams contained significant amounts of formaldehyde. Most of the new ones don't, but their use is still restricted in some states. Pricing tends to be close within any one local area.

If you need a higher R-value, you have to go to a so-called specialty pre-insulated block. This consists of a specially molded block and a special EPS insert designed to fit into it. Figure 5-8 contains diagrams. These blocks usually have two webs instead of three, and the webs are cut down. This limits the amount of heat loss.

The advertised R-values of specialty inserts are generally 10–15 for an 8-inch block. The manufacturers determined some of these with a guarded-hot-box test, and some with a parallel path or isothermal planes calculation.

Specialty pre-insulated block sometimes gets resistance at first from masons. At corners you have to use a standard block and inserts, which lowers the R-value there. For some brands, you also have to switch to standard inserts wherever there's vertical reinforcing.

The price quotes we got on specialty pre-insulated block varied a lot. In one case, we got prices from two different plants 150 miles apart for the same brand of pre-insulation. One plant was apparently adding $1.35 to the cost of the block for the insulating feature, and the other was adding $3.00. It's probably worth calling around to compare.

So far as we know, there's currently no way to make a half-high block with any of the brands of specialty pre-insulated block. Most of the other popular architectural block styles are available, however.

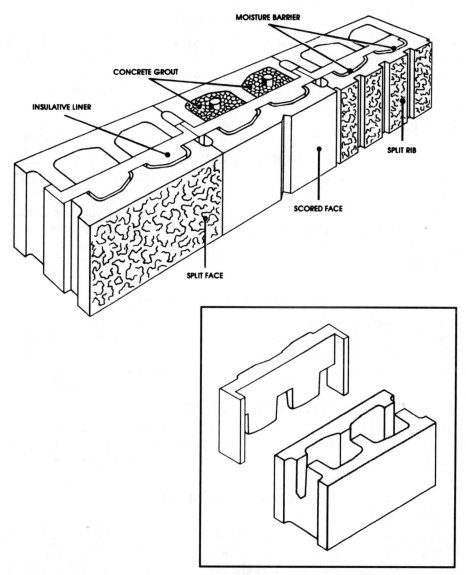

MOISTURE BARRIER

CONCRETE GROUT

INSULATIVE LINER

SPLIT RIB

SCORED FACE

SPLIT FACE

5-8 *Diagrams of specialty pre-insulated blocks and inserts.* Insul Block Corp.

To find specialty pre-insulated block, call local block plants or contact the insert manufacturers directly to get the names of plants that carry their brand. They're under "Concrete block pre-insulation" in the back of the book.

The total thermal mass of a cavity-insulated wall is about the same as an interior-insulated wall, minus the small additional mass of sheetrock. Almost any cavity-insulated wall will meet the mass of 6 Btu/square foot/degree Fahrenheit required for preferential R-value treatment in building codes.

Thermal mass

In a cavity-insulated wall, roughly half of the mass (about 4 Btu/sf/degree) is inside the insulation envelope. Homeowners we talked with verified that they felt more comfortable in their homes than in others they had been in. We have no data on solar or load-leveling applications using cavity-insulated walls. Buyers looking for these things usually picked other systems that offer even more interior mass.

Water resistance

As with interior-insulated block, we had no complaints from occupants about water penetrating walls. We were told stories about this happening, but these appeared to involve older houses that used old materials or weren't repainted on schedule. Because cavity-insulated walls frequently have no sheetrock inside, probably any water entering the block stands a greater chance of becoming visible.

Maintenance required

Other than painting or sealing the outside, builders and owners told us they never have to do anything significant to keep their walls or finishes in shape. With this system, that applies to the interior surface as well. Scuffs or pressure on the concrete surface have less effect than on sheetrock.

Disaster resistance

Cavity-insulated houses should have similar disaster resistance characteristics to interior-insulated homes. One builder speculated that they might have superior performance in fire because they have a concrete surface on both sides.

Product availability

Most of the popular block products are widely available. Slump block is available from only some plants, concentrated mostly in the Southwest and on the West Coast. And specialty pre-insulated block is not currently available in a half-high unit. This is an important consideration in the Northeast and Midwest because you need a specialty pre-insulated to get sufficient R-value, and half-high is one of the more popular looks there.

The different types of cavity insulation have spotty availability in some areas. Most cities we surveyed had a couple of pour-in-place installers and a couple of foam installers nearby. That's not a lot. It makes us think that there could be some rural areas where the nearest installer is a long distance away. The cities seem to have one or two block plants within shipping range that carry standard pre-insulated block, but rural availability might vary. We've heard that the insert manufacturers can sometimes set up a local plant to produce it if they have your order in hand.

Labor availability

Installers of pour-in-place and foam insulations are not always plentiful. Several builders also stressed the importance of finding good ones and, in some cases, keeping on top of them. The quality of their work varies. The problems we heard about were failure to fill some cavities completely and failure to show up when the insulation work was scheduled. Late arrival is a problem if you're installing the insulation just before each bond beam is grouted.

We heard that masons are sometimes skittish to bid on laying pre-insulated block because they don't know how long it will take them. But others are less concerned about trying something new. And the ones who've used it before say it goes about as fast and easily as ordinary block.

Getting an adobe look on slump block is a specialized skill. Some masons and stucco crews in the Southwest do it, but you have to ask around to find them.

Building department approval

We didn't find anyone who had to do anything unusual to get building department approval. Everything in this system is already allowed by nearly every code, although some things might be in parts of the code that are more commonly applied to commercial buildings.

Required calendar time

There was a huge range on construction times. The most basic houses went up even a tad faster than interior-insulated block houses because there were no furring strips, interior studs, sheetrock or insulation to put inside the walls.

For the more expensive custom homes, the feeling was that work took longer than frame construction. The builders told us that the schedule could get extended by the sum of a lot of little coordination delays and miscues. Sometimes work had to wait on the insulation crew. Sometimes it had to stop while forgotten electrical lines were put into the block. But we found it impossible to separate the impact of these things from the dozens of other factors that extend the schedule of any custom home.

Crew coordination

Although these houses can go routinely, there were a couple of coordination issues that could take close supervision. One is the installation of insulation, already discussed. This was most clearly a concern when the wall is reinforced on a 4-x-4 grid and pour-in-place, or foam insulation is to be placed without putting holes in the wall. In that case, the insulation crew needs to arrive at every 4 feet of wall height, just after the vertical reinforcing cells are filled with grout and just before the bond beam is poured. In practice this means that you want both grout and insulation crews on the site at the same time. When there's a mixup, it might mean grouting the vertical cells then sending the grout crew home until the insulation is poured.

The other coordination issue is installation of utility lines in the walls. The builder has to make sure that the masons know what the plumbers and electricians want. This can be as simple as having detailed plans and hiring a careful masonry crew. Or it can require some complex timing of masonry and utility crews, as was the case on some of the expensive houses.

Making connections

The popular connection methods for the cavity-insulated system are the same as for the interior-insulated system. The only extra consideration has to do with windows. Usually there was little or no trim used. This was either to save cost or to leave as much block exposed as possible. To do that, you need to pick one of the window types that connects directly to the block without a buck.

Utilities installation

In affordable housing projects, there was a strong preference for putting as many plumbing and electrical lines in the interior walls as possible. We're told it's simpler and cheaper. In that market segment, the buyers are not highly sensitive to floor layout, so they might accept somewhat unconventional arrangements if necessary to route utilities through inside walls.

In the quality projects, some of the more difficult rooms get furring strips and sheetrock anyway. This was common in kitchens. But overall there were more plumbing and wires in the block. The buyers were less willing to compromise on the placement of utilities, and cost was less of a consideration.

There were two distinct philosophies to putting utility lines inside block. One was to have all plumbing and electrical lines marked clearly on the plans and let the masons provide for them. If a piece of pipe needs to run through, the builder makes sure the masons have the pipe. When they come to the point in the wall where the pipe goes, they cut holes in the block where it emerges, set the pipe in position, and lay the block around it.

For electrical lines, the procedure is similar. They cut holes in the block, including rectangles where the boxes will go, and lay plastic conduit into the wall. The electrician comes in later, threads the wire through the conduit, and attaches the boxes. Sometimes you can omit the conduit and thread the bare cable through the cavities, but some codes require conduit. It's inexpensive anyway, and some crews feel more comfortable using it.

The other philosophy is to have the utility trades on-site as the wall is built. The plumber or electrician shows the masons where to make their holes, cuts the pipe or cable, and installs it. This is obviously more expensive, but it might be required by crews that adhere to detailed work rules. And there are few mistakes this way.

Change flexibility

The costs and logistics of making changes in the structure are all the same as with interior-insulated mortared block. The one additional consideration is that more care has to be taken with the work for expensive homes because the appearance of the block is so important.

Manufacturer support

The one unique product in this system is the insulation. Most of the manufacturers are small companies with representatives in only some parts of the country. The builders we talked to dealt more with the local installation crews and block plants. The amount of service they provided varied. But we also heard that the manufacturers could be helpful. They give out names of local installers or plants, answer questions, and try to solve problems by phone.

Other considerations

As with other block systems, occupants of cavity-insulated houses claim their interiors were unusually well protected from sound outside. There were no particular cleanliness issues with the system. The only unique product it uses is insulation. You might have to clean up some stray bits of this.

One slightly more serious consideration could arise when foam crews spill their ingredients or when the foam leaks out of the walls. Some of the chemicals stain block. This is unimportant if the walls are later painted, but it's potentially a big problem if the house is to have an exposed architectural surface. In one house, the masons had to knock out and replace some stained block. Fortunately, it was in an unreinforced part of the wall. The solutions are to make sure you have a good foam crew, go over the situation with them, and make sure the masons put in tight mortar joints all around to prevent leakage.

Exterior-insulated mortared block

Exterior-insulated mortared block is ideal for the energy market segment. The system puts large amounts of thermal mass inside the insulation envelope, and the insulation can be increased to very high levels with modest increases in cost and effort.

The flip side is that exterior-insulated homes are more expensive than most frame construction. Depending on how far you want to go, the cost can be $2000–$6000 higher than 2-x-6 fiberglass-insulated frame for a typical 2000-square-foot home. But that buys a level of comfort and energy efficiency that ordinary frame doesn't have and probably couldn't get without an even higher extra expense and a pretty awkward design. So for the people who want really high energy performance, the exterior-insulated system can be a cost-effective way to go.

The one novel ingredient to the exterior-insulated homes is a special breed of insulation system that attaches to the outside of the block. Most of these go loosely under the name exterior insulation and finish system (EIFS). Some people call them "Dryvit systems" because Dryvit was the first brand that got a lot of attention. They've been available for years, but until now mostly in commercial construction.

Exterior-insulated construction has attracted buyers in northern climates because of their solar heating capabilities. In the southern states, they're attracting interest mostly from utilities that want to encourage nighttime air conditioning. The components are available all over.

Construction begins with a conventional block wall. It starts to look different afterward when the insulation crew comes in. The insulation consists of foam board, up to 4 inches thick, that's attached to the outside surface of the block. It's covered with a finish material. This is usually some variant on stucco, but there are now some fancy alternatives, too.

There are a couple of distinct variations on the exterior-insulated home. The simpler and less expensive version leaves the block exposed on the interior. In this case heat transfer between the wall and inside air is direct. Often the builders specify an architectural block with the finish side faced inward to get a distinctive interior.

The other variation is more expensive but offers more extensive energy features. It sometimes goes by the name Xenwall, which was coined by the inventor. Increasingly it's being called OPTM, which stands for off-peak thermal mass. This name was coined by the North Florida Concrete and Products Association, which is promoting the system for its ability to be cooled at night. In an OPTM house, every cavity of the block walls is filled with grout whether it contains rebar or not. This nearly doubles the thermal mass. The inside walls get furring strips and sheetrock. The HVAC is always air, and it's distributed not through separate ducts but between the block and sheetrock. In this way, the heat or air conditioning gets stored in the block, which passes it back to the living space gradually and continuously.

How the air circulates in an OPTM house appears to get worked out project-by-project. The original design by the inventor called for metal furring strips that don't quite reach to the top and bottom of the wall and force the air to go up and down the block surface. A more recent version includes a trench in the slab foundation that carries air to the bottom of all the exterior walls, where it rises straight up between the strips, out vents along the ceiling, and back into a central return in the interior.

In cold climates the OPTM system is used to blow warm air (from solar collectors or rooms with a lot of south-facing windows) over the block. In warm climates it cools the block at night with the AC system.

In the north some buyers interested in energy efficiency are specifying OPTM homes. According to a major developer in the South that built an OPTM house in cooperation with a utility, for sizable sales of these houses to develop, the power companies will probably have to promote them and offer subsidies.

Labor and materials' costs

The cost depends a lot on what version of the system is used. For the simplest version, without extra grout, or any furring or sheetrock, we got quotes around $3000 extra to build a house of about 2000 square feet of living space. The builders giving us these numbers were comparing to 2-x-6 construction with fiberglass batt insulation and a clapboard or quality stucco siding. For the more elaborate OPTM houses, the quotes we got were that they would cost $4000–$6000 more than frame for a medium-size home.

Table 5-4 lists approximate costs per square foot taken from actual quotes. The costs of the basic block wall are about the same as with other mortared systems. For a simple exterior-insulated house, the additional costs are the EIFS and (usually) some sort of architectural block on the inside. Builders that used EIFSs told us they add $5.00–$7.00 per square foot of wall area, depending on the thickness of the foam used and local labor rates. However, some builders got this down to $4.00 by putting together a conventional stucco and foam board off the shelf instead of using a brand-name product. A split-face block pointed inward and careful mortar work might add about $.80. The total for the simple system would thus be around $9.50–$10.50, depending on the amount of reinforcing.

Table 5-4.
Representative Costs of Exterior-Insulated Walls

Item		Cost per Square Foot
Unreinforced wall		
Materials		$1.30
Labor		1.90
Reinforcing		0–1.00
None	$ 0	
8' × 8' grid	.75	
4' × 4' grid	1.00	
Extra grouting		$ 0–.75
None	$ 0	
Full	$ .75	
Exterior insulation and finish		5.50
Interior finish		$ .80–2.20
Split-face surface	$ .80	
Furring, sheetrock, paint, sealing	2.20	
Total		$9.50–12.65

A full OPTM house has the additional costs of full grouting, plus sheetrocking and sealing joints and channeling the HVAC through the walls. Grouting costs are under a dollar a square foot if you're reinforcing anyway. Most of the cost is the extra concrete. But if you're figuring the cost on top of an unreinforced wall, it will likely be more than $.75 in the table. The usual cost of interior sheetrock (including furring and paint) appears to go up about $.50 when you do the extra work to seal it and circulate the HVAC behind it. This all leads to a total cost for an OPTM house of about $11.65–$12.65, depending on the reinforcing.

Learning costs

Builders who produced houses of the simplest type reported start-up costs and homework similar to builders working on interior-insulated homes. Those who had built with block before usually had nothing to learn about except the exterior insulation and finishing. Their background work consisted mostly of tracking down good EIFS crews through word of mouth. Those who took on an exterior-insulated house as their first block project had to learn about the usual details of block in addition to the insulation and finish.

Building an OPTM house for the first time is more of an adventure. Although the inventor of the system recommends certain construction details, there's still a lot of room for discretion in how the HVAC equipment is set up and how the air is channeled behind the walls. The builders we interviewed invented a lot of their own components and design features, things like a trench in the foundation to distribute air to the base of all the exterior walls and curved pieces of sheet metal to channel air around curves.

The OPTM builders estimated they spent $2000–$3000 extra in materials and labor on their first house. This means they spent as much as $9000 over the cost of 2-x-6 construction the first time. Most of the extra on the first house was from slow work by crews who were unfamiliar with the new tasks they were performing or for time spent rigging up some new assembly no one had used before.

This does not include the extra research, design, and thinking time that the OPTM builders put in. One independent general contractor estimated his personal time in these activities at 160 hours. The construction supervisors of a major developer couldn't give us an exact number of hours, but said that two to three people in the office had it on their minds in odd moments and in their spare time for months.

Buyer reaction

We couldn't find that people had a strong reaction to the materials themselves. Some buyers came to the builder requesting an energy-efficient house, and they got enthusiastic when they learned what this system could do. Others had heard of the system and looked for someone to build it for them.

Contractors who built one of these houses on speculation reported that the average buyer couldn't understand the unique features from a description. A builder in the Northeast had an architect do drawings of the wall section and, as he put it, "nobody cared." The first impression depended more from the house's appearance (usually a stucco exterior and sheetrock or exposed block inside) and how these struck personal taste.

But people who spent more time in one of the houses said they began to notice its higher comfort level and quietness. This impressed most, and helped them understand the explanations of the unique energy features.

Exterior and interior finishes

Builders either used a prepackaged EIFS or put together their exterior from connectors, foam, lath, and siding purchased separately. Buying separate components was more common in the southern half of the country. The builders fastened off-the-shelf foam board to the block with any of a number of things. Some used a long masonry screw with a large washer to keep the screw from pulling through the foam. Some preferred a wood screw and plastic anchor because they thought the plastic would allow less heat to transfer out of the house. The metal lath went up at the same time. It was laid up against the foam, and the washers pinned it into place.

Thermomass Corporation supplies plastic pins that you can implant in the mortar of the block before it dries. You then push the foam panels between the projecting pins and screw the lath onto the pins' ends. Figure 5-9 shows a picture. A builder that used these pins said a lot of care was necessary to make sure they stuck out to exactly the required distance. Otherwise, an even stucco job would be difficult. But he felt they worked well. The manufacturer is included in the appendix under "Exterior insulation and finish products."

5-9 *A block wall with pins sunk in mortar joints to hold foam board and wire lath.*

After the foam and lath are up, the stucco can go on just as it does over sheet foam and lath on a frame house.

In the Midwest and Northeast, the finish was more often part of a complete exterior insulation and finish system. The report we got from the builders and sub-contractors was that these systems are better formulated to withstand the freeze-thaw cycles of their region than PC stucco on lath. They're available from several manufacturers. They include all the necessary materials, except maybe the foam. Figure 5-10 contains a picture of the parts of one system. They attach the foam with a plastic connector or an adhesive. They use a PM or PB surface material. First comes one coat of material, then a fiberglass mesh for strength (instead of a metal lath), then more material. The final coat usually has color premixed into it.

5-10 *Layers of an exterior insulation and finish system: 1-substrate, 2-adhesive (also used to attach foam board to substrate), 3-molded EPS foam board, 4-fiberglass mesh, 5-finish coat.* Corev America, Inc.

The EIFS manufacturers are now offering optional premium finishes, with things like decorative stone embedded in the stucco. Keystone Systems of Connecticut supplies prefinished foam panels with a wide variety of looks. The panels are glued directly to the block and sealed around the edges. All of these companies are listed under "Exterior insulation and finish product manufacturers" in the appendix.

One builder used a 4-inch-thick decorative block wall outside the foam to form a double-wythe wall. Experienced masons know how to do this, but it adds considerable cost.

Although they'd never done it, some builders agreed you could probably attach any of the lap or plastic sidings to the outside. You would simply attach furring strips outside the foam instead of metal lath.

The interior finish is either exposed block or sheetrock. All of the OPTM houses used sheetrock inside. Builders of some of the simpler houses glued sheetrock directly to block inside. This let them use ordinary gray block and keep good thermal contact with the inside air.

Design flexibility

The structural material in this system is nonproprietary block, with the design flexibility characteristics of that material already described. Stucco and the EIFS can cover a surface of almost any shape, so they don't restrict design appreciably. For inexpensive trim and details, a lot of builders glued extra pieces of foam on before stuccoing.

R-value and energy efficiency

The R-value of the wall depends mostly on the type and thickness of foam used. The builders we interviewed favored 3-inch or 4-inch sheets of one of the EPS foams or polyurethane, for a total insulation R-value of 15–20. The block, the air gap, and sheetrock (if any) add perhaps 2–5 to the R-value, for a total of 17–25.

A major advantage of the insulation in this system is that it's nearly unbroken. The connectors pass through, but if they're plastic the heat losses should be small. There's an occasional penetration by things like outdoor spigots, but these are also minor. We don't have guarded-hot-box data on complete wall sections, but logic suggests they'll be close to the figures given. One bigger potential for heat leakage is at the roofline. Our builders typically covered the top of the block and top plate with insulation.

The mass-corrected R-values of one of these walls will clearly be higher than their other R-values. If the block is ungrouted, they should be about 50 percent more in very warm climates and 5 percent more in very cold ones (as discussed in chapter 4). The mass corrections would be even higher for the grout-filled OPTM walls. They have a weight around 60 pounds per square foot and a thermal mass that we estimate at 15 Btu/sf/degree Fahrenheit. But we don't know of anyone who's done the mass correction calculation.

Obviously, any of these walls will qualify for reduced insulation requirements under the codes that allow them. However, that's probably not very important because their buyers prefer a lot of insulation, not the minimum requirement.

Thermal mass

The thermal mass of these walls is virtually all inside the insulation. That comes to about 8 Btu/sf/degree for the empty block and 15 for the fully grouted.

The occupants heat an exterior-insulated house in upstate New York entirely with cordwood from a central fireplace. The electric backup heat never comes on. The house uses ungrouted block and is just under 2000 square feet (living space).

The occupant of an OPTM house in Minnesota said that on nights when the low temperature was in the 20s, the furnace wouldn't come on until morning. The heat stored in the block kept the house reasonably comfortable all night. This house also had more south-facing glass than an average home, and when the days were sunny and in the 40s, the furnace never came on. The heat through the windows carried the house through to the sunshine of the next day. Even in the coldest month of 1993–94, which had nighttime temperatures of 30 below, the gas bill of the 2200-square-foot house was $85.

The OPTM owners told us that in hot weather the house stays comfortable without AC so long as the nights are in the sixties. Temperatures above the mid-seventies day and night lead to an uncomfortable indoors in a little less than a week. Those that didn't have or didn't turn on AC described the indoors as, "better than other houses without air conditioning, but not as cool as houses with AC."

The latest OPTM houses in the South rely solely on air conditioning in the night and morning. This cools the block enough to keep the house comfortable through the afternoon and evening, even on hot summer days. Some owners, however, complained of high humidity. The block is supposed to absorb moisture until the AC dries it at night. However, in houses that channeled the air only a short distance over the block, this might not always have worked effectively.

Everyone, North and South, agreed that the temperature of their houses was more comfortable than the neighbor's frame house. The temperature was highly even. In the winter there were no "cold spots" in the walls. The HVAC clicked on and off rarely.

Water resistance

The barrier to outside water is the siding outside the foam. No one has reported any problems yet, but some of the users of stucco or EIFS had some concerns.

Engineers specializing in the subject claim that in commercial buildings, systems that use some form of stucco over foam have had a higher rate of problems than most sidings. The problems usually stem from some sort of break in the hard stucco shell. It wears thin, develops cracks, or gets dented. Water has sometimes penetrated through the insulation layer.

On the other hand, the builder of a 6-year-old exterior-insulated house in New York State claims that nothing has come through, and the finish is still sound. So it appears to be possible to get durable results with these systems.

The specialists we talked with claim that a lot of the problems in commercial construction appear to be the result of the field installation. When you're dealing with a material that's supposed to be troweled on to a thickness of less than half an inch, skimping by just an eighth can make a big difference. In addition, sealant or flashing is called for at some joints, and this isn't always applied as carefully as it needs to be. The message is to make it a priority to find a crew that knows its materials and works conscientiously.

Maintenance required

The only maintenance issue that arose was concern about the stucco-on-foam and EIFS exteriors. It's difficult for us to comment on this because we have no cases of problems in the homes we found. The manufacturers claim that their new materials are more durable, and that the same materials can be used to patch any problem areas.

Most EIFSs have color mixed into the top coat. Because of the coat's thickness, some builders have projected that this coloring should last at least 10 years before any painting is necessary.

Disaster resistance

Exterior-insulated houses should behave in fire, wind, or earthquake about the same way an interior-insulated house behaves. Some builders speculated that the greater mass of the OPTM house would help it survive the uplift forces of wind better than block walls with ordinary reinforcing. In addition, once you've decided to go with fully grouted walls, the extra cost of more reinforcing is extremely small. It's just the cost of the rebar. So you can go up to almost any wall strength relatively inexpensively.

Product availability

The only unusual product used for this system is the EIFS. The manufacturers ship the parts of their systems across the country. There are dealers in just about any large urban area, and one or two within striking distance of most rural areas. Some builders that bought directly allowed a few days for delivery. But more of them relied on the stucco subs, who are now mostly familiar with the EIFSs.

Labor availability

The block work for these houses required the same skills as interior-insulated and cavity-insulated houses. The issues about mason availability were therefore similar.

The OPTM builders said that there's no obvious trade to do the furring and air channeling work inside the walls. The builders themselves invented a lot of the details, so they usually did the work themselves and got some of their more creative and broadly skilled subs to help.

The other labor issue that cropped up was to get a good crew for exterior insulation and stucco. The best advice our builders could give us was to use word of mouth and check out people's past work. If you use a prepackaged EIFS, you might want to ask the manufacturer for an installer in your area. The manufacturers appeared to give out the names of trained people who do quality work. This makes sense because they wouldn't want their products to get a bad name.

Building department approval

Structurally, the exterior-insulated houses simply had to meet the usual block requirements. They all did, and some of the OPTM houses far exceeded code, so officials were quickly satisfied.

All the OPTM builders also told us that their codes didn't allow an exposed cable in the space between block and sheetrock because it's technically an air duct. They solved this problem in two different ways. Some ran all the wiring in the exterior walls in metal conduit. (Plastic was not sufficient for an air duct, according to their codes.) Some used their metal furring strips. With some designs of steel strips, you can form a sort of metal tube by putting two together.

Required calendar time

The simpler exterior-insulated houses took about the same time to erect as any other block house. The general opinion was that exterior foam with a stucco finish took maybe half a day more than interior insulation and an exterior stucco finish on a medium-size home. On the other hand, the exterior work didn't hold up the rest of the job because other things could go on at the same time.

The special air-handling provisions of the OPTM houses added calendar time. The builder with the most elaborate air system estimated the extra time to install the extra features at two weeks.

Crew coordination

In virtually every house, the electrical lines need to be in conduit. They're either in the block (when there's no sheetrock or it's glued on) or they're in an air duct (in an OPTM house). The only exception was when the cable ran through metal channel furring strips. In addition, in an OPTM house the builders felt it was necessary to seal up any possible openings that might let air escape from the space between the sheetrock and the block.

These things are not part of the customary work of most subcontractors, so the builder has to keep track of who's going to do what. There are also the usual considerations of putting openings in the structure in the right place from the start, so block doesn't have to be cut or filled in later.

Making connections

There were few unique connection details. Floors, roofs, interior walls, windows, doors, and interior fixtures and cabinets went on the same way they do with other types of nonproprietary block. For windows, the crews had to form a sill over 2–4 inches of foam as well as the block. They did this by cutting the foam at an angle and putting a thicker layer of finish material over it or by laying an extrawide preformed sill.

The OPTM builders used a few tricks to seal connection points against air leakage. They fastened 2-x-4 plates to the block where interior walls were to be attached, put the sheetrock all around, and only then nailed their interior walls to the plates through the sheetrock. They put sill sealer under their top plates and adhesive around their bucks. In OPTM houses, most of the interior fixtures get attached to the metal furring strips. Sheet metal screws were the favored fastener for that.

Utilities installation

As mentioned above, electrical cable almost always went into conduit. In an OPTM house, the electrician could install his own conduit during rough-in because he had an open wall to work with, just as with frame. Or, when it was available, he could use the channel in the metal furring strips. In the other houses, someone put the conduit in according to plan as the block was being laid. Likewise, plumbing went the way it goes into a conventional frame wall or a conventional block wall, depending on whether the house was an OPTM or not.

Change flexibility

The methods and costs of structural changes are similar to the other nonproprietary mortared block systems. None of the builders of OPTM houses made a change in their exterior walls, but they pointed out that it would be much more difficult after the walls were grouted. For either an OPTM or simpler system, changes made after the exterior materials are up require cutting or building up the foam insulation and getting the stucco crew out to cover up the bare spots.

Manufacturer support

As with other systems, local block plants provided a lot of support on the block parts of the structure. The EIFS manufacturers give help over the phone and send product literature: installation manuals, specifications, and so on. The builders usually relied on their local EIFS product distributors or installers for advice. Some of them preferred to deal with someone who carried several brands and let him decide exactly which product would be best for the job. In the Southwest the stucco seemed to be treated more as a commodity, with the builder rarely getting into the details. This is perhaps because the crews are so accustomed to working with stucco on foam from frame construction.

Other considerations

The owners of all of these houses say they're extremely quiet. The most extreme report we got was from the owner of the OPTM house with a 4-inch block veneer outside the foam. It's located in a flight path. He claims that his neighbors sometimes complain of being awakened by loud airplanes or thunderstorms during the night. But they've quit asking him what he thought of the ruckus because they've learned he never hears it.

The owners of OPTM houses pointed out that they rarely had internal drafts from their HVAC systems. Because the air flowed behind the walls, the interior was especially still. They found that the lack of moving air added to the comfort.

OPTM owners also had one useful tip: don't put the spot for the kitchen garbage next to an outside wall. The heating within the wall makes the food waste spoil quickly. The builders also decided that in later houses they wouldn't put kitchen cabinets on the exterior walls for similar reasons. These things weren't mentioned in the South, however, probably because the walls are mostly cooled instead of heated.

The Hebel Wall System

Technically, the Hebel Wall System consists of blocks held together with mortar. But that's about where its similarity to mortared block construction ends. It uses almost all new materials: new concrete, new mortar, new finish. The system is more expensive than conventional frame, but it also produces houses with features that fit the quality market segment pretty much right down the line.

The block is not molded; it's cut from an entirely different form of concrete. It's called *aerated autoclaved concrete* (AAC), or sometimes *autoclaved cellular concrete* (ACC). AAC is made with all fine materials, nothing coarser than finely ground sand. Included is an expansion agent. Because of this agent, after the concrete is mixed it rises like bread dough. And like bread, the finished product is filled with countless small air holes. In fact, it's about 80 percent air. The factory cuts it into precise units.

The material has been used in Europe since before World War II. By the 1980s it had spread to just about every part of the world except North America. It's been built into millions of residences worldwide, sometimes in the form of blocks and sometimes in the form of larger panels. There are even prereinforced floor and roof panels.

Recently, the world's largest AAC manufacturer (Hebel of Germany) set up American subsidiaries to sell the product here. It's now sold in the United States through Hebel Southeast, which is in Atlanta. Currently they're offering the block system, but in the future they expect to sell panels as well. Because they ship from Georgia, nearly all of the construction has been in the Southeast. But Hebel plans to build plants and make product available across North America over time.

Compared with conventional concrete, AAC has some unusual characteristics valuable for home building. It insulates well. It cuts, drills, and takes fasteners a lot like lumber. And it's light.

The standard Hebel block used for exterior walls is 8 inches thick. It's solid; there are no cavities. Its height and length (a nominal 10 inches by 25 inches) give it about twice the face area of the conventional 8-x-8-x-16 block. Yet, at 37 pounds, it's only slightly heavier.

The first course is set and leveled on the foundation with standard mortar. Thereafter you use a special "thin-bed" mortar supplied by Hebel. It mixes in a bucket with water and a wire stirrer that fits on a standard drill. You lay it with a special trowel—also supplied by Hebel—that has guides so you can put down a precise 2–3 millimeter bed on the tops and sides of the blocks. Blocks set onto the mortar are tapped secure with a rubber mallet. The blocks are almost self-leveling. They're cut to precise dimensions, and the mortar bed is so thin that it leaves little room for error. Just aligning the block with its neighbors nearly trues it. A level reveals remaining deviations; you adjust with taps from the mallet.

It's relatively easy to form odd angles and openings of any dimensions because the block cuts so easily. A standard hand or circular saw for wood works fine. Figure 5-11 shows some bays made with the block. Hebel offers a special coarse-toothed handsaw that cuts quickly. Some claim it cuts as fast through an entire Hebel block as you can cut a 2-x-4 with a handsaw. Some crews also put a band saw on site for fast, precise cutting. Where cut blocks are needed, the crew usually measures and cuts them individually before laying.

5-11 *Home of stacked and mortared Hebel block before stuccoing.*

For bond beams and lintels, there's a U-block. It gets reinforced with the standard rebar and grout procedure. Bond beams are formed at the top of each floor. When a lintel falls on the course of block below the bond beam, you can make holes in the bottom of the bond beam U-block directly above the lintel. In that way the grout poured into the bond beam fills the lintel as well. Figure 5-12 shows the block ready for such a pour. For lintels lower down in the wall, the grout must be poured immediately after the lintel course is laid. Lintels are held in place for grouting by temporary braces made of 2-x lumber.

5-12 *Bond beam of Hebel U-block cut at bottom so that grout will also enter the lintel below.*

The Hebel wall system uses less vertical reinforcing than conventional block. The manufacturer explains that this is possible because the Hebel mortar provides a stronger connection than standard mortar joints. In the houses we encountered (which all had walls designed by a structural engineer), there was one rebar at each corner. However, there were usually none alongside window openings. The manufacturer said that in areas with strong wind codes, the engineer might specify one additional vertical rebar every 15 feet or so.

Vertical rebar are ungrouted. One dowel embedded in the foundation marks each spot for vertical reinforcing. The blocks laid at those locations are drilled with a hole of an inch or more, and they're threaded over the dowel. Figure 5-13 shows some drilled block. Matching holes are drilled all the way up, forming a narrow cavity up to the bond beam. Before the bond beam pour, another rebar is dropped down the cavity to overlap the dowel. The two rebar are connected by welding, and the upper bar is locked into place on top by the grout of the bond beam. An alternative reinforcing method uses threaded rod instead of standard rebar and connects successive rods with nuts instead of welding.

Any chips or holes in the block are filled in with a special repair mortar (also supplied by Hebel) after the wall is up. Uneven spots can be sanded out with a coarse paper; Hebel offers a large, flat sanding block.

Interior walls can also be formed of Hebel block. A 4-inch-thick block is stacked the same way the exterior walls are. There are also preformed lintels to go over doorways. Some interior walls appear in Fig. 5-14. Because of the

5-13 *Hebel blocks cut and drilled on-site to form extra U-block (left) and rebar cavities (right).*

5-14 *Interior walls of 4-inch Hebel block with preformed sills above the doorways.*

weight of the block, the interior Hebel walls are an option only over slabs or commercial-strength floors. There's also a 10-inch Hebel block available.

To date, almost all the Hebel homes in the United States have been built in urban areas of Georgia and in various locations in northern Florida. All of the builders we interviewed about the system operated in those places.

Labor and materials' costs '

Some builders suggested that a one-story house of about 2000 square feet of living space would cost about $2000 more to build with Hebel exterior walls than with 2-x-4 frame walls. Others said it would cost significantly more than that, but didn't give exact numbers. Getting reliable figures was difficult because few builders have used the system on more than one house yet. Some builders pointed out that they had saved some money because the insulation and thermal mass of the system had let them put in smaller air conditioning units. One estimate we got for a large (4000-square-foot) home was that it was possible to use a 5.5-ton instead of a 6.5-ton unit, saving about $1100 up front.

The unit costs we got were $7.50-$8.00 per square foot of gross area for the complete wall. This includes all labor for the mortared and reinforced block, all openings and cuts for utilities, an exterior finish consisting of a high-quality stucco supplied by Hebel, and plaster applied directly to the block on the interior.

It's also possible to use conventional stucco outside and sheetrock inside at some savings. But, as discussed below, Hebel feels that the other products produce a higher-quality result that's more consistent with the expectations of the people who buy their system. And most of the builders chose to use these finishes. Quotes we got for Hebel's stucco were about $1.80 per square foot, including materials and labor. One plasterer said that he charges $1.00 per square foot total to do the inside of a Hebel wall.

It's generally not possible to break out the costs of different components of the system. Builders who want to use the system on a particular house supply the plans, whereon Hebel figures out the package of materials needed to build it and gives a quote for the total.

Learning costs

Some builders felt confident with the system after one house, and all of them said they had the kinks out after two. The builders who quoted numbers said that they spent an extra $2000 or so on their first house because of their lack of experience with the system. Bear in mind that they were building predominantly large (3000–4000-square-foot) homes.

There were a few things mentioned that contributed extra cost on the first job. One was a failure to get the first course of block precisely straight and level. Most people working with conventional block don't position any one course of block extremely precisely because they don't have to; the mortar joint can take up the slack in succeeding courses. With Hebel, if the first course isn't level, you have to use conventional mortar again on the second course to correct. That slows things down.

A failure of the block crew to lay a perfectly flat wall was another start-up problem. If there are uneven spots, the time and materials of the finish crews go up.

There are also some differences in the tasks and sequencing of the crews. You'll read about the major ones under "Making connections" and "Utility lines."

They're logical and not difficult, but people work a tad slower until they get used to them.

And finally, the wall crew simply needs one or two jobs to get really proficient with the new tools and materials. One builder told us that some masons he hired never really adapted to the new methods. When things weren't going well, he got a different, more flexible crew to finish the work.

As for the extra time the builder spends personally to plan out the Hebel house, we were told that on the first house it runs 20–30 percent more than usual. Part of this is a one-day training session that Hebel gives to all the builders using its system. There's a slightly different session to teach the subcontractors who install and finish the walls.

Buyer reaction

The general reaction of U.S. buyers seems to be curiosity. When given a sample of the material they see that it's different. AAC is light in color (it's almost pure white) and weight. Houses going up attract attention because of their unusual appearance—a thick, solid white wall with rectangular seams and precise corners. People tend not to think of the material as "concrete." This might be helpful if their old images are negative, but it might be necessary to spend more time explaining the material and construction methods. The manufacturer does a lot of this educational work with interested buyers, however.

There's a strong positive reaction from European immigrants. They're more familiar with the material from their homelands, and many of the early customers are people from Europe who heard that it was now being offered in the United States. Some talked as though they consider it to be the standard for construction, and consider anything less to be shoddy. One German told us, "I just can't stand the thought of a house made out of wood."

Exterior and interior finishes

Like concrete in general, AAC provides a good surface for stucco. No lath or surface preparation is necessary. Hebel has developed its own lightweight stucco that goes on with one coat. It matches a lot of the characteristics of the block: density, rates of expansion, and so on. This means that the block and stucco will expand and contract together, minimizing cracking. Hebel supplies it and recommends it. It's possible to use just about any other stucco as well.

Stucco is the siding chosen by nearly all builders so far. But in principle, just about any other siding could go on, too. The ease of sinking fasteners into the block might make alternative sidings interesting for builders willing to experiment. Vinyl or aluminum can be screwed in directly; furring strips could go on relatively easily to provide a backing for nailed materials like clapboard.

Inside, Hebel recommends direct plastering or, as a second choice, furred-out or glued-on sheetrock. The builders we spoke with used plaster. It was a tad more expensive, but they agreed with the manufacturer that it produces a higher-quality result.

Design flexibility

The system has not yet been used in basements in the United States that we know of. It's been approved by the SBCCI for the purpose. However, the cost would be high and, in the words of one of the manufacturer's people, "In a basement you don't really take advantage of its special properties." The manufacturer claims the system can go to four or five stories in height. The builders to date have generally gone up one or two.

Odd angles and unusual openings are relatively easy to make with Hebel block. The judgment of our builders was that they could make them with the same cost and degree of effort as they can with frame.

A non-90-degree angle involves cutting each block at the angle point before laying it. The cutting goes quickly on a band saw or with a Hebel-supplied guide and handsaw. One subcontractor estimated that an odd-angle corner would add $150–$200 in total cost, although he didn't always break them out in his bill.

For an odd-shaped window opening, builders would sometimes put up the wall and then mark and cut out the hole in place. This can be done with a standard circular saw if the shape consists of a series of straight lines, or something like a reciprocating saw if it curves. Cutting in place is easy and ensures proper alignment of the blocks. If the top of the opening is straight, it can be capped with U-block so a lintel is formed. If it isn't, a lintel can be made just above the top point of the opening, and the block below the lintel can simply hang from it. This is one advantage of the light weight of the block and high strength of the mortar.

R-value and energy efficiency

Hebel's AAC has been tested in the laboratory with a guarded hot plate (which performs virtually the same type of test as a guarded hot box). The ACC has an R-value of about 1.1 per inch. This gives the 8-inch block a guarded-hot-box R-value of about 9.

Because the structural material is the insulation, there are very few complete insulation breaks. The mortar joints are about one-third the thickness of conventional block's joints. And because the block is bigger, there's less of them. At vertical reinforcing points, there's only a minor hole in the center of the wall. At bond beams and lintels, the U-block still has about 1½ inches of insulating AAC on either side of the grout.

The manufacturer has had engineers calculate a mass-corrected version of the R-value of their wall. It's a little different from the mass-corrected R-values we generally report in this book because the engineers also took into account air infiltration. They estimated that because a Hebel wall has so few gaps in it, a Hebel house has about 2.6 air changes per hour. This compares with about 7 air changes per hour for a standard frame wall. Naturally, this lowers energy consumption. So taking into account both the thermal mass and the lower air infiltration rate, the engineers calculated that energy consumption in a Hebel house in Orlando, Florida would be lower than that of a stud wall with R-30 insulation.

Thermal mass

Judging by the weight of the block, the thermal mass of an 8-inch Hebel wall should be about 4 Btu/sf/degree Fahrenheit. It's difficult to say whether this mass is inside the insulation or outside because the thermal mass and the insulation are formed by the same material. Builders claim that when they stand inside Hebel houses without air conditioning on warm days, there's a noticeable thermal mass effect. The cool block keeps the interior cooler. Presumably, it can also store heat that's useful in passive solar and store coolness that's useful in load-leveling applications.

Water resistance

Testing indicates that the Hebel wall's resistance to water penetration is outstanding. The manufacturer says that in the wind-driven rain test, a wall without any exterior finish didn't soak through at all for 100 hours. Presumably this would be even longer if there were a finish on the block.

Maintenance required

So far, the builders report no callbacks and no maintenance on the Hebel walls they've built. Probably the houses with painted exteriors will require repainting about as often as conventional stucco does—every six years or so.

Disaster resistance

A standard 8-inch Hebel wall survived a fire test for four hours without a failure of any kind. The company projects it might be able to get as much as an eight-hour rating, but it hasn't performed the longer tests yet. This would be nearly unheard of for residential exterior walls.

Hebel houses are currently being built to Florida wind codes with one vertical rebar about every 15 feet and at the corners. According to the manufacturer, if requirements were to go even higher, they could be achieved with a few more rebar.

There hasn't yet been any earthquake testing in the United States. However, the manufacturer is optimistic about the product's earthquake resistance because of the strength of the mortar, the possibility of adding almost any amount of reinforcing to the wall, and the advantages of light weight in shaking.

Product availability

Hebel Southeast currently imports the product from Hebel of Germany and ships it from its offices south of Atlanta. They're willing to send it almost anywhere for the incremental shipping cost. For a recent house built in Connecticut, this was about $700. However, according to our builders, the availability of support from Hebel Southeast is more of a consideration. For their first couple of houses, the builders relied on the extensive support that Hebel provides. This would probably not be as readily available in other regions.

Hebel currently plans to license at least two plants in North America to manufacture the products starting in 1995. One would be in South Georgia and the other in Monterrey, Mexico. The Mexican plant could economically ship to parts of Texas and New Mexico. There are plans to build more plants to supply all regions of the United States and North America over time.

Labor availability

Hebel Southeast provides one-day training sessions for crews that install the wall system. It offers to recommend a good local crew that they've trained, or to train your favorite crew free of charge. They won't sell the system unless there's a trained crew.

From the reports we got, a good portion of the conventional masons take to the system fine. According to a Hebel sales representative, "With about half of them, on the second day with the stuff they have an "Aha" experience. They realize all the things they can do with the system and they get enthusiastic." But others don't adapt as well. We're told that they try to use the Hebel products the same way they use the conventional ones, which isn't always right. For example, many will try to use the standard Hebel mortar (instead of the stiffer repair mortar) to fill in nicks in the wall. The result can be slower, lower-quality work. An example is in Fig. 5-18.

More recently, Hebel has located tradespeople from Europe in the Atlanta area that have taken well to the system. Many are familiar with it or similar products, and they generally seem to think it is a good material, worth taking some extra effort for.

Surprisingly, carpenters take well to installing the system. The block cutting and wall truing are right up their alley, while some of the traditional masonry skills like mortaring have been simplified. So it matches the carpenter's capabilities well.

The manufacturer has reported to us that once the Hebel "masons" get good, they can do a house with only themselves and several less-skilled laborers under their supervision.

Building department approval

The Hebel wall system now has an SBCCI evaluation report. However, the company pays to have an engineer analyze and stamp every set of plans for which it provides product. First the company reviews the plans. Hebel might suggest changes, then has the engineering work done and deals with the local building department, if necessary. The builder is spared a lot of the legwork. As far as we know, no Hebel house has ever failed to get building department approval. Hebel met with the building departments in advance to explain their system, which probably helped smooth the way.

The builders we spoke with said there is little for the inspectors to look at in the exterior wall. Some wanted to see the rebar before the bond beam was grouted. If the inspectors need to see the vertical rebar, they can do so through the lookouts cut in the block where the vertical bar and foundation dowels meet. These lookouts have to be cut to connect the two rebar, anyway. One is shown in Fig. 5-15. The bond beam rebar are, of course, visible before grouting from the top.

5-15 *Lookout cut into Hebel block with rebar ready for welding.*

Required calendar time

For the large houses they were dealing with, our builders said that it typically takes a week to construct the exterior walls. This includes placing the rebar, grouting the bond beams, and routing out channels for the plumbing and electrical. The walls go up this fast with a crew of four to five. The routing takes a single person about half a day.

Because the mortar sets fast, the roof structure can go on two days after the wall is laid. So the roof, plumbing, electrical, and windows can all go in during the second week. Finishing the exterior can also occur in the second week, or later if preferred.

Interior plastering takes perhaps another week. The builders working in cold weather said that they could not lay block in sub-freezing temperatures. The mortar froze. However, it sets fast, so it should be safe to install above freezing even if the temperature will fall below overnight. The manufacturer said that no special measures are needed for extremely hot or dry weather other than standard masonry requirements.

Crew coordination

The sequence of crews in the Hebel wall is nearly the same as with frame. One big exception is with electrical work. After the wall is up, the block is routed out where electrical lines will go, and the block is cut where boxes will go. Figure 5-16 shows this operation. The material cuts easily, and in theory electricians could do this work. But they don't normally carry the tools. So on the jobs we heard about, the electrician came in and marked where everything would go directly on the wall. Then a member of the wall crew came back and did the routing and cutting. The routing is best done with a bit that makes a slight lip in front so the cable rests securely.

5-16 *Routing channels in a Hebel wall for electrical lines.*

Plumbing is handled about the same way as with conventional block. When plumbing lines stick up out of the foundation, the masons cut their block and work around it as necessary. Because they have the entire 8 inches of wall to work with, even vent stacks are accommodated relatively easily. Furring out the wall was never necessary in the houses we saw. Pipes that go in after the wall is up can use a routing procedure, just like the electrical.

After the lines are in and inspected, the plaster crew fills in the cut and routed voids with Hebel repair mortar. A picture of this is in Fig. 5-17. Over larger channels, like vent stacks, a layer of fiberglass mesh goes on over the cement for added strength. If the wall will receive sheetrock, there's no need to fill in the channels.

5-17 *Troweling repair mortar over recess cut in Hebel wall for a vent stack.*

If the house has an intermediate floor deck, there's also some back-and-forth between the wall crew and the carpenters. When the wall is up to the second floor, the carpenters come in and put on the deck, then the wall crew comes back to continue the wall.

Making connections

Hebel Southeast provides tables of concrete connectors that it recommends for use with the system. These are things such as cut nails and wood screws with plastic anchors. But there are two other optional fasteners that are very handy because they don't require any predrilling.

One is the Hebel nail. It comes in lengths of 2, 4, and 6 inches, with a head about the size of a large bolt head and a thick, square shank. We nailed in a couple. It goes into the AAC with about the same effort as a twelve-penny nail goes into a solid piece of lumber. Nailing it in is very true, without any noticeable chipping or irregularities along the way. It goes through lumber as well, so it can be used to attach 2-x to the wall.

The other option, which some builders told us about, is to put a screw directly into the block without any predrilling. The manufacturer recommends using anchors for heavy loads. But our builders reported using plain screws for fixtures as heavy as cabinets and for attaching windows. One builder favored long (3½ inches or longer) drywall screws because he found shorter screws and wood screws didn't hold well enough. In addition, builders who used plain screws reduced the speed of their drivers because turning too far stripped out the block.

The fastener chosen to connect interior walls to the block was usually the Hebel nail. If the interior walls were made of the 4-inch block, the connection was with an L-shaped metal brace, fastened to each wall with the nail or screws. The entire assembly needed to be countersunk so plaster or sheetrock could go over smoothly.

Joists for intermediate floors went in pockets. After pouring the bond beam for the first floor, the carpenters set the joists in place, bearing about 4 inches on the walls. Then the masons cut blocks to a depth of 4 inches as necessary to build around the joists and continued building up the wall.

Windows and doors can be attached just as is done with frame. The flange is screwed directly to the exterior surface of the wall, then sealed and covered with siding. We assume that the various methods of attaching masonry windows used with conventional block could all work with Hebel, too. Another method sometimes used involved a flangeless window with a screw-through frame. The workers screwed the frame to FHA straps (a 6–8-inch metal plate generally used by plumbers), then set the window and screwed the straps to the block to secure the window. One builder found this awkward and time-consuming, however, and decided to try another method the next go-round. Fixtures can go up with screws, with or without anchors, or sometimes with the shorter Hebel nails.

Utilities installation

Channels for the electrical and plumbing lines are cut into the block as noted previously. The builders generally found that this was an easy procedure. Wherever a penetration in the wall was necessary, drilling with a conventional wood bit did the trick.

Change flexibility

The ease of making later changes is one place where Hebel shines. Openings and wall sections can be cut out easily wherever there's no rebar in place, and there's generally not much rebar.

We saw where one builder had cut out a window opening late. The wall crew had missed it on the plans and filled in the section where it was to go. The builder caught it before the lintel course, marked it, and cut it out in place by using his portable circular saw on each side of the wall. The two cuts met in the middle. The whole job took perhaps two hours. A picture of the result is in Fig. 5-18.

Another builder told us about an owner who decided he wanted a window moved a foot to the left. The crew cut out one side of the opening and filled in the other. There was no need to sawtooth the block on the fill-in side, and there were no rebar alongside the opening to move. The lintel was long enough that it still spanned the entire opening by a big enough margin to be effective. Otherwise the crew would have had to knock out part of the wall to rebuild or extend the lintel.

5-18 *Window opening cut out of a standing Hebel wall, with cutout portion on the ground below. Notice that masons have incorrectly used block mortar instead of repair mortar to fill nicks in the block, causing the drips on the wall.*

Manufacturer support

The builders all tell us that Hebel Southeast was involved all through the planning and construction process, helping at every step. In some cases a buyer had contacted Hebel, and Hebel asked one of the builders they're familiar with whether he wanted to take on the job. In other cases, the builder got a client interested in the system and then came to Hebel.

In either case Hebel will meet with the buyers to go over technical specifics, costs, and so on, and get them comfortable with the system. They take the plans to do a takeoff and prepare a price for the materials package. And they help the builder prepare a final price for the buyer. If the buyer signs a contract on the job, Hebel has a structural engineer do specifications and put a seal on the final plans. It provides training sessions to local building departments and keeps contact with officials to ensure acceptance of the system and of the individual sets of plans.

Hebel will recommend qualified crews or train new ones. During construction, builders ask for advice and get it. According to one, "When we started this house, Mike (a Hebel representative) told me I could do anything with this stuff that I can do with frame. I said, "Okay, we'll see." Now I keep coming back to him saying, "How do we do this? How do we do that?" And he keeps coming up with a way to do it."

Other considerations

The manufacturer had preliminary tests done that indicate an 8-inch Hebel wall has an STC of 40 without finishes. Hebel Southeast expects the final tests (to be completed in late 1994) to show a much higher STC.

Cutting the AAC produces a fine white dust. According to the trades, there's not very much of it, and it causes little mess or problem. The block is very resistant to cracking or shattering. A whack with a hammer produces only a dent. The same goes for dropping the block onto a hard surface.

Further information

Hebel Southeast
4000 Cumberland Parkway
Suite 100B
Atlanta, GA 30339
(404) 344-2897

The Integra Wall System

Although masons stack up Integra block much like conventional block, and the finished wall looks about the same afterward, there are big differences inside. The block has only one web, and that one is cut down. The block is pictured in Fig. 5-19. The vertical reinforcing is not conventional rebar. It's "tension rods"—steel

5-19 *Basic Integra block (bottom), end block (top), with half, corner, sill, and half-high blocks (left, top to bottom).*

rods, threaded at both ends and attached at the top and bottom of the wall and tightened down, so there's no grout. And the cavity between the inner and outer face shells of the block is filled with a specially formulated foam. The result is an exceptionally strong and well-insulated masonry wall for a cost about equal to 2-×-6 frame.

The Integra Wall System is the invention of Superlite Block of Phoenix, Arizona. Superlite is a large producer of concrete masonry. In the late 1970s, the arrival of the wood-frame-and-stucco system from California started to take a big share of the Arizona market. Much of the reason was that it was more awkward and expensive to build a block wall that had as much resistance to earthquake and as much insulation as frame has.

Superlite's engineers suggested using post-tensioning. In a post-tensioned wall, steel reinforcing rods run through the middle of the block and hold it together the way the string holds together the pearls of a necklace. The rod connects to the block only at the top and bottom. At the top of the wall, it sticks up through a special block that's solid except for a narrow slot. A picture of it is in Fig. 5-20. A metal plate, a washer, and a nut go over the end of the rod and onto the slotted block and tighten the rod so it compresses the blocks down the entire height of the wall. If side or uplift forces push on the wall, the pull of the tension rods holds it in place.

This type of groutless reinforcing and the special Integra block keep the breaks in the insulation from grout and block webs limited. Yet Integra Wall Sys-

5-20 *Top course block with recess to hold metal plate.*

tem walls are well enough reinforced to satisfy Zone 4 earthquake requirements.

The system can be installed largely by existing trades with little special instruction. One of Superlite's goals was to make it easy for ordinary masons to understand and use. The two steps that are novel—the tensioning and insulating—are handled by a special crew that Superlite sends to the building site.

The properties of the Integra Wall System make it well-suited to the quality market. But it has recently begun to get sales in what we would consider the value segment as well. At early 1994 prices of lumber, it matched 2-x-6 home prices in some projects we surveyed.

Construction to date has been almost exclusively in Arizona and Southern California. Most of the builders have been large ones that used the system in high-volume production developments. But according to Superlite, a rising number of custom homes are using it as well.

Until recently, Superlite was the sole seller of Integra Wall System components and the reinforcing and insulating service. But Superlite is partly owned by a company called Oldcastle, which owns other block producers around the country. Some of the other Oldcastle-owned companies are working with Superlite to pick up the Integra Wall System and offer it in their areas. As of press time it's also available around Salt Lake City, Utah, and might be available soon in North Carolina and around Washington, D.C.

Other than the unorthodox reinforcing and insulating, the logistics of Integra block installation are similar to conventional mortared block construction. The foundation crew has the one added responsibility of embedding a special threaded bolt into the poured concrete wherever a reinforcing rod is to go. Figure 5-21 shows one of these. Rods normally go every 4 feet o.c., plus at corners and within 1 foot of either side of an opening. The exact reinforcing is determined by the engineers of Superlite, who adjust and stamp all plans for builders using the system.

5-21 *Special anchor bolt embedded in foundation, with plastic cap to keep debris out.*

After the foundation is poured, the masonry crew starts by screwing one end of the tension rods into the bolts in the foundation. A photo of a foundation with rods in place is in Fig. 5-22. Then the masons build their walls. On most houses they use a 6-inch-wide block, although an 8-inch is also available. Sliding the blocks around the rods is easy because the blocks have only one center web.

5-22 *Slab with tension rods.*

In addition to the standard block with one cut-down web and the nearly solid unit that goes on top of the wall, there are special shapes for corners and sills, pictured in Fig. 5-19. Steel lintels are usually used above openings. As with a conventional block wall, there are also full and half block with closed ends to form the jambs of openings.

When the wall is up, one of Superlite's Integra Wall System crews puts a steel plate, washer, and nut over each rod, and tightens it down to a precise tension, then injects an expanding polyurethane foam into the cavity from above and cuts off the excess. If the builder wants, the crew also attaches 2-x top plates for an extra charge. The plates bolt onto the remaining ends of the tension rods. Figures 5-23 and 5-24 show these operations. At this point, finishing can proceed just about as it can for any other block wall.

Labor and materials' costs

Builders in the Phoenix area agreed that when the price of lumber rose to its early 1994 levels, the cost of building a home out of Integra block edged even

5-23 *Integra Wall System crew pouring insulation into a wall.*

5-24 *Integra Wall System crew tightening down reinforcing rods and attaching top plates.*

with 2-×-6 frame. Their comparison was to homes with stucco exteriors and sheetrock inside, which are the dominant finishes in their area.

Table 5-5 includes per-square-foot costs we received from Arizona builders in the summer of 1994. In the volume production operations where we saw the system used, the cost of the structure plus insulation could go as low as $3.50

(for a perfectly rectangular floor plan) or as high as $4.00 (when there were a lot of corners). The bill for the Integra Wall System crew includes installation of wooden top plates.

Table 5-5.
Representative Costs of Integra Wall System Exterior Walls[1]

Item	Cost per Square Foot
Masonry bill (materials and labor for block and rod installation)	$2.50
Integra crew bill (insulation, rod tensioning, and top plates)	$1.25
Subtotal, structure and insulation	$3.75
Exterior stucco and paint	.90
Interior furring and sheetrock	.90
Interior painting	.50
Total, including finishes	$6.05

[1]The costs listed here are in the middle of the range of quotes we received from builders reporting their actual project costs.

Learning costs

The builders insisted that the extra time and money they spent on their first Integra block houses was virtually zero. Superlite trained the foundation and masonry crews as necessary and handled almost all the unusual tasks, as we've noted. So for the general contractor, it went like any other block project. One builder told us that when he sent his first Integra Wall System home out for bid to masonry crews, he learned that all the crews he solicited had already built with the system before. He had to teach them nothing.

Obviously this is largely a result of the careful groundwork Superlite has done in designing the system for traditional trades, training local crews, and plugging its own crew into the construction cycle. Using the system might involve more homework for builders who try to use it in a totally new region.

Buyer reaction

The Arizona buyers we talked with were informed about the wall system from the developers selling it, and the buyers liked it. We heard from several people that concrete in general is considered a high-quality product in that area because of its solidity, insect resistance, and coolness. Upon hearing the description by the developers' sales staffs, buyers were further impressed by the apparent sturdiness and energy efficiency of the Integra Wall System.

One of the production builders summed up the appeal of Integra block pretty neatly. To paraphrase him, his company focused on people retiring to Arizona with a little more money than the average buyer. They wanted a quality home they could feel good about living in for the rest of their lives. If he used frame, many of them might buy from him anyway. But the Integra Wall System showed that the company was a cut above the competition in construction quality, just as it was in other details. The system was one more factor that convinced buyers they were dealing with a quality organization and made them comfortable with their decision to buy.

The reaction in other parts of the country will probably be somewhat different simply because impressions of concrete are different. We don't have first-hand information because so few Integra block houses have been built outside the Southwest.

Exterior and interior finishes

The most common exterior finish is painted PC stucco. Superlite also offers the block in a range of architectural finishes. It estimates that the typical architectural wall with reinforcing and insulation costs the builder about $5.00 per square foot for the structure and insulation. If the block is a half-high, this can go to $6.00. Furring and sheetrock are additional. However, we hasten to point out that there will be no additional cost for exterior finish materials. That's about a dollar a square foot that doesn't have to be paid for paint and stucco.

When exposed block is used, the masonry crew has to make sure they have consistently tight mortar joints. Where they don't, the foam will squeeze out of the block. Excess foam can be cut away, but it will discolor the block irreversibly. Needless to say, that's probably not acceptable for someone who's paying a premium for an architectural finish.

Although we never heard of it being done, in principle you should be able to put any of the lapped sidings (clapboard, shingle, vinyl, aluminum) on the exterior as well. You would just add furring strips first, as with conventional block. Inside, furring strips and sheetrock were virtually always used. A typical furred interior appears in Fig. 5-25.

Design flexibility

The system can be used for basements, but these are rare in the Southwest. According to the manufacturer, when this is done you use 8-inch block, more frequent reinforcing, and conventional damp proofing on the outside.

Some houses have been built with two stories above grade, using the 6-inch block. Superlite claims that it's possible to go higher, but the costs go up about as they do for conventional block because scaffolding and materials-handling can slow work. Multiple stories involve some unusual details because the reinforcing rod must pass through a bond beam at the top of the first floor. But Superlite has worked this out and provides details to builders.

The builders we interviewed stuck with 90-degree angles and rectangular wall openings. Superlite can supply special 45-degree angle block. Other odd angles would be more difficult. The block would have to be cut, as with conventional masonry, and any gaps would have to be filled with mortar to keep

5-25 *Furred interior of an Integra house, with some excess foam cut flush with the wall.*

the foam insulation from pushing through. Odd-shaped openings are normally handled by making a larger, rectangular opening and filling it in with frame.

As with conventional block, construction is faster and cheaper if the walls and openings are designed on the 8-inch module. All the builders we interviewed strove for this. They had a lot of help from Superlite; the company helps to revise the builder's plans if necessary.

R-value and energy efficiency

The Superlite crew fills the walls with a polyurethane foam that the company had formulated by a plastics company just for the task. The crew squirts two liquids into the wall, where they react to foam up and fill the cavities. The foam allegedly squeezes into even small nooks and crannies. Because it's a polyurethane, it has no formaldehyde and can't give off formaldehyde gas.

The manufacturer calculates the parallel-path R-value of the insulated wall to be R-15 for the 6-inch block and R-20 for the 8-inch block. This takes into account the insulation break formed by the web in the standard block. Other insulation breaks occur at the window sills, door and window jambs and lintels, and at some extra webs that are in the corner blocks. In a multistory house, the grouted bond beams between floors would also be insulation breaks.

Owners of Integra Wall System homes report that their air conditioning comes on later in the year than the AC of neighbors in 2-x-6 frame houses, and their energy bills are lower. The company claims AC savings of about 12 percent versus conventional frame construction and says that this is a conservative estimate. One owner said that the AC in her neighbor's wood frame house comes on the first day the outside temperature goes over 90 degrees. At the time we spoke, Phoenix had just experienced its first two days of the summer, back to back, with daytime temperatures of more than 100 degrees. The AC in her Integra block home with conventional double-pane windows had not come on yet, but the interior was still comfortable. The six-inch Integra block has, by our estimation, a thermal mass of about 6–7 Btu/sf/degree Fahrenheit. This is sufficient to qualify for additional R-value credit under most energy codes.

Thermal mass

About half of the thermal mass of the completed wall—3 to 3.5 Btu/sf/degree Fahrenheit—is located inside the insulation envelope. Superlite also offers a version of the block with a thicker inside face shell. That face shell is about 3½ inches instead of 1½ inches, which should logically give it a thermal mass inside of about 7.5 Btu/sf/degree. It's intended for utility-sponsored load-leveling applications, but it could serve as solar heat storage as well.

Water resistance

The manufacturer believes that the urethane foam acts as a moisture barrier, although this hasn't been tested yet. The foam's moisture resistance would be in addition to the barrier provided by the paint and stucco. The homeowners we talked with claimed not to have noticed any water damage or moisture penetration. The builders said they had not had any callbacks because of water getting through the walls. The homes were all in the relatively dry Southwest, so it's hard to generalize to other regions of the country. But our preliminary evidence shows no water problems.

Maintenance required

The builders we spoke with were involved in major development projects, so they routinely took responsibility for any problems with the structures they built several years after construction. They claimed that the only problems with Integra Wall System homes were occasional hairline cracks in the stucco. They said such cracks occurred in perhaps 10–20 percent of their Integra block homes, and that this is less frequent than it is with their stucco over wood frame. Their solution was to use a patching material and paint over it. Other than this, we heard of no need for maintenance on Integra block walls beyond the periodic repainting that all painted walls must go through.

Disaster resistance

A wall of insulated 6-inch Integra block achieved a fire rating of 1¾ hours in testing. After that time, the cold side exceeded the allowable temperature limits, but the wall was still structurally intact. The manufacturer claims that twice it's learned of a fire alongside the completed wall of an Integra Wall System home. In both cases the wall itself survived intact. The foam insulation was singed along the top where it was exposed, but apparently was unharmed inside the block.

The Integra Wall System homes built to date have been designed by Superlite's engineering staff to withstand 80-mile-per-hour winds. This meets local codes. Superlite believes that a bond beam might be required to meet the code requirements of high-wind areas like Florida. They think this might be done by ending the reinforcing rods at the course below the top course, then putting a conventional bond beam on top of that.

The system also meets Zone 4 earthquake requirements with its standard reinforcing. If a builder wants to increase the number of tension rods to increase the margin of safety or protect the house from unusually harsh conditions, the cost is small. Each 8-foot rod costs about $3, and the incremental labor cost of installing one more is small enough that no one could make a precise quote on it for us.

Product availability

Currently, Superlite supplies Integra block to builders all over Arizona and in parts of Southern California from its plants in Phoenix and Tucson. At greater distances, they say, shipping costs make the block less economically attractive.

Recently, Amcor Block of Salt Lake City started to offer the Integra Wall System. Amcor is another Oldcastle-owned block producer. It has promoted the system mostly for use in commercial construction in the Salt Lake area.

Adams Products Company, another Oldcastle block company headquartered in Morrisville, North Carolina, is attempting to get SBCCI code approval so that it can introduce the system in the Southeast. A third company, Betco Block, is also planning to market it in the Washington, D.C. area. According to Superlite, the intention ultimately is to sell it from block companies around the country.

Labor availability

Except for Superlite's reinforcing and insulating crew, construction of Integra Wall System homes requires only standard trades. The concrete crew needs some instruction in how to place the anchor bolt for the reinforcing rods, and the masons need some directions in how to use the block. Superlite has a field supervisor to do this, however, and none of the builders we talked with had to become involved.

The masons we talked with felt that there was little new they had to learn to use the system. Their biggest early problem, they claim, was just learning how to handle the odd-shaped unit. With only one web, the block can break fairly easily. It took the crews a little practice to adjust their habits so that they could cut down on their breakage.

The ready availability of masons familiar with the system probably results from Superlite spending an unusual amount of effort in working with them.

Finding trained crews might be more difficult in areas where Integra block hasn't been available as long as in Arizona.

Building department approval

The Integra Wall System has a complete ICBO evaluation report (No. 4845P). But in addition, Superlite has its engineers adjust and stamp all builders' house plans before submission to the local building department. Superlite has also educated many of the building departments in their area about the system and helps guide plans through the approval process.

The builders said they had to do little work of their own to get approval for their houses. They submitted their plans to Superlite, and the company took care of most building department concerns.

Required calendar time

The builders said that Integra block took about as long to erect as conventional block. Some said it goes slightly faster because some steps related to reinforcing are eliminated.

For an average ranch house on a slab in high production, erecting the block could be as long as three days. But the really experienced crews that used seven people did the job in a day and a half. The reinforcing inspection is eliminated. Superlite is authorized in most of the areas it sells to do the inspection itself. There's also no grouting. After the mortar has set for 24 hours, the Integra Wall System crew can insert foam, tension the rods, attach the top plates (if assigned to do so), and inspect in 1–2 hours. The roof framing can start right after this. Furring strips inside took another 1–2 hours. So about 3½ days pass in total. At this point the wall is ready for stucco and sheetrock.

In contrast, for a similarly sized house of lumber the builders allotted 1 day for framing the exterior walls, and another for sheathing with foam. That's two days before the structure is ready for stucco and sheetrock, but the stucco on frame takes as much as a half day longer because of the need for lath.

We've never heard of Integra block being installed in extremely cold weather, but the rules that govern conventional masonry probably apply. The manufacturer thinks that the insulation could proceed on schedule even in the cold. Superlite has different blends of polyurethane for different temperatures, including versions that work in unusually cold and hot weather. Block work is also delayed by extremely rainy weather, but Superlite claims that insulation can proceed even if the block is wet. The expansion of the foam allegedly pushes the water out through the block.

Crew coordination

Everyone said that there's no extra work to coordinating crews with the system. It's even slightly easier than conventional block because there's no grout to work in. The reinforcing and insulation is all that's unusual. That's handled by Superlite, and it comes neatly between the masons and other crews. The only provision the masons ordinarily have to make for later crews is to work around some plumbing coming out of the foundation, as is done in conventional block work. If the block will be exposed, there's one additional consideration. The mortar work must be especially tight so that no foam comes through.

Making connections

Most of the connection details are the same as for conventional block. If necessary you can put conventional 6-inch blocks in the wall—such as U-blocks to form bond beams, ordinary block that can be grouted to anchor J-bolts, and so on.

To connect floor decks at intermediate floors, the conventional block methods are used. Using conventional block as necessary, the crew forms pockets or sinks J-bolts into a bond beam to hold a ledger. The roof framing simply attaches to the top plate.

The windows we saw used are the variety that's anchored into the mortar with metal clips and fits flush in the jambs of the opening. The masons slide a window in place when they're part way up its opening, then continue laying around the window. Figure 5-26 contains a photo. But it also appears possible to use any of the other types of window connections common to block construction.

5-26 _Window held into place with clips embedded in the mortar joints._

The builders we surveyed attached their furring strips with masonry nails from pneumatic guns. The sheetrock and interior fixtures and trim then connected directly to the strips. We heard of one builder who glued sheetrock directly to the block, then put most of the utilities and fixtures on interior walls. The interior walls get connected to the block with cut nails or more masonry nails.

Utilities installation

The crews got most of their plumbing and electrical between the furring strips. The only exceptions were some vents and large water lines. These the masons worked around as they went up, just as with conventional masonry. Everyone agreed that running the utilities was no more difficult than with interior-insulated mortared block.

Change flexibility

Probably because they ran production operations, none of our builders ever had to make a change in any of their structures once they were up. According to the manufacturer, before the foam is in, cutting and moving things goes about the same as with conventional block. In some ways it's easier because there's no grout to cut. When there is a change, Superlite redoes its engineering and submits that to the building department.

After the foam is in, however, knocking things out gets tougher. The manufacturer says that the foam "acts like a cement." A mason told us that trying to make a hole with a sledgehammer was extremely slow because the wall would just bounce back after he hit it. He preferred heavy wrecking equipment.

Manufacturer support

The production builders we talked with agreed that Superlite was excellent at taking care of their problems. They felt that because of the support, there's almost no cost or inconvenience to switching over to the Integra Wall System for a builder who's familiar with conventional block.

They submitted their plans to Superlite, which revised them as necessary to fit the 8-inch module and performed the engineering work. Superlite would also work with the building department if there were any questions about the adequacy of the product or the engineering. A site supervisor from the company was available to train masons without the builder having to be involved. According to one builder, once you're underway, "They send sales reps to your site if you need it. If you wanted to bring them out to help figure out something unusual, they'd put a sales rep on your site as long as necessary for you to get comfortable with it."

Praise for Superlite's support was so great that we tried to track down builders who no longer use the product to see if we could find anyone who had any criticism. But those builders had stopped using the Integra Wall System only because they decided it didn't fit the market they were trying to appeal to. They agreed that Superlite had given them great support and made it easy to use the system. We did find one custom builder who got interested in using Integra block, but then changed his mind. He felt that Superlite's support is more designed to meet the needs of the large builder who will be using the product in high-volume developments.

Other considerations

We have no data on the sound resistance of an Integra block wall. We suppose that the STC is similar to a conventional block wall, although the foam might add to the sound-dampening effect.

The Integra block construction sites we saw were unusually neat. After insulation, the Superlite crew cuts protruding bits of foam away with a saw. They're small and appear harmless, and can be collected easily.

Further information

Superlite Block
4100 W. Turney
Phoenix, AZ 85019-3327
(602) 352-3500

Sun Block

Sun Block is extremely simple. It consists of a lightweight, high-insulating, 8-inch block that can be cut and fastened to about as easily as wood. But the logistics of putting the house up are almost identical to putting up a house of standard block. It's just a little easier, and the final wall insulates better and repels water, on top of having nearly all of the usual advantages of concrete.

Sun Block's special properties come from its materials. Unlike a conventional block, it's made of a mixture of cement, fine sands, and EPS beads. They're mixed throughout the cement, which makes the final product lighter and better-insulating than standard concrete. It also cuts, drills, and screws with standard wood tools.

Sun Block is the invention of Sparfil International, Incorporated. The company also makes a very different, mortarless block system (the Sparfil Wall System II) using the same material, which we describe in the next chapter. Sparfil sells Sun Block only in Florida through its offices there.

Masons install Sun Block the same way they do conventional block, including reinforcing. If special blocks are needed, you can slip in ones made of ordinary concrete or cut a Sun Block as necessary to fit the job. In general, anything that involves cutting the block goes about as quickly and easily as it does with lumber.

The buyers have always been classic Quality segment buyers—people who are willing to pay a little more for features like better insulation, durability, and unconventional architectural details. Because all the Sun Block homes to date have been built in Florida, that's where our information comes from.

Labor and materials' costs

The Florida builders claimed that the cost difference between conventional masonry and Sun Block is just about equal to the higher cost of the block itself. In early 1994 a Sun Block cost about $1.00 more, which translates into about $1.13 per square foot extra. On a typical 2000-square-foot ranch, this works out to something around $1500, depending on the exact layout.

If the block's R-value is sufficient for the buyers' needs, you can save some money by eliminating the interior insulation. This is under a dime per square foot for the builders we spoke with. Their normal insulation is ¾-inch fiberglass. You can also eliminate the furring strips and glue sheetrock directly to the block. In that case utility lines are routed inside the block or interior walls. One builder also claimed to have saved about $500 on his AC equipment. Because of the block's R-value, he was able to install smaller AC units.

The builders also claim to have saved a little masonry-crew time because the block is light and cuts easily. Builders have reported that occasionally in the past the block was a little uneven in size, and this slowed work down slightly.

There are some savings over conventional block because of lower waste. The Sun Block is less prone to breaking. One builder claimed his crews lost 2–5 percent of its conventional block (about 30–75 units for a midsize ranch) from breakage, but rarely lost any Sun Block. You also save time and money connecting walls and fixtures because standard drywall screws do the job without drilling.

Learning costs

One job seemed to be enough for everyone to figure the product out. The masons had to learn that it's tougher to knock out a hole with a masonry hammer. The material just bounces back for the first few hits, so you have to keep tapping longer or switch to a saw. But after that, masons generally like working with the product. It's lighter—about 25 pounds per unit instead of 37—so they get less tired. And it cuts more easily.

The later crews liked it because it's easier to connect to than regular concrete, and it cuts with a carbide blade in a standard wood saw. They picked up on all they had to know after a few hours of working with it.

The builders said they spent a few more hours on their first house because they weren't familiar with the product, but it was a small amount of time they didn't keep track of. They sometimes had to push a new masonry crew not to increase charges for use of the unfamiliar product. They also had to learn just what the product could do; some parts of the wall that they would normally have framed they could now leave to the masons.

Buyer reaction

One developer kept a sample of the block in his display room. He said that if people could see the product, they appreciated it more. His experience and that of the other builders we talked with were about the same. It impressed people interested in more energy efficiency than provided by the standard home.

Somewhere around 5 percent of all of the developer's buyers ended up specifying the product for their homes. He felt this was more than enough to justify offering it because it helped him attract customers he might not have gotten otherwise, and because it required little extra effort to offer and use. Another, smaller custom builder we talked with specialized in energy-efficient homes. His customers didn't usually know about the product, but when he explained it to them, many liked it and wanted it put in their houses.

The buyers of Sun Block homes were almost all comparing it to interior-insulated mortared block. In the central Florida area, there is little frame home building, and block is highly regarded by the market.

Exterior and interior finishes

The exterior was always some form of stucco. It's popular in Florida and it goes directly on Sun Block just as it does on other concrete surfaces. Because Sun Block can be screwed with wood fasteners, furring strips for other sidings should go up readily. Sheetrock was always used inside.

Design flexibility

Our builders always built on a slab, so we have no examples of basements made of Sun Block. But according to the manufacturer, it has been used for walls that were partially below grade.

Houses as tall as three stories have been built. Since Sun Block doesn't have quite the load-bearing strength of standard block, such high walls might require a few more cavities to be filled with grout to pick up the load. The Florida Sparfil representatives can give help on this for a particular design.

Most design features involving cutting went almost as easily with Sun Block as with lumber. This includes odd angles, bays, and odd window openings. Some builders made unusual openings by stacking up the block and cutting it in place. However, the opening still usually had to be flat above for the lintel. The lintel was a course of grout-reinforced block, which is hard to form in a curved or zig-zag pattern.

R-value and energy efficiency

The plain Sun Block tests in a guarded hot box at R-8. In the final wall, the ⅜-inch mortar joint around each block is uninsulated, however. Any conventional concrete blocks that are used are also uninsulated. But if you're willing to cut Sun Blocks into U-block, halves, and other specially shaped units, you don't have to use many conventional blocks. Reinforcing does not eliminate the insulation. The grout is still surrounded by the special insulating concrete.

The manufacturer calculates the thermal mass of a Sun Block wall at 6.2 Btu/sf/degree, and the weight at about 28 pounds/sf. This is enough to qualify for special insulation credit under most codes. Sparfil has also calculated the mass-corrected R-value of Sun Block walls for the central Florida area. It's about R-20.

Thermal mass

Sun Block is one of those tricky systems in which the thermal mass and the insulation are intertwined. It's hard to say the mass is inside the insulation envelope exactly, or to say that it isn't. Since heat could get into the cement portion of the block from the surface, and could come back out into the air, presumably the block would add to the energy storage of a solar house if no interior insulation were put over it. It should also contribute to the coolness storage of load-leveling houses. But it would probably be a stretch to assume that all 6.2 Btu/sf/degree of the total thermal mass would be available, since the part near the exterior of the wall would take pretty long for the heat or coolness from the inside to reach.

We have no examples of passive solar or load-leveling houses using the block, so we can't give field reports of its performance in those applications. On the other hand, most of the buyers were clearly interested in the high R-value and potential for energy savings from the block. They reported air conditioning savings on the order of 30 percent, compared with their friends who had interior-insulated block with ¾-inch fiberglass insulation. They also claimed that the walls had a generally "cooler" feel.

Water resistance

The manufacturer thinks that the EPS beads might add some water repellence to the Sun Block. We saw a wall built partly from Sun Block and partly from standard block that had gone through a heavy rain. The standard block was dark with water, while the Sun Block stood out because it was about as light as when it was dry.

In the houses we heard about, the block wasn't counted on to provide water resistance, however. The builders put up stucco and paint for that. None of the buyers complained about water getting inside through the exterior walls.

Maintenance required

Maintenance on Sun Block was no different than for conventional block. Depending on the variety of stucco used, there might be some hairline cracks that can be patched. Painting is a little rarer than with wood-based sidings. We heard of no problems or callbacks to do with the concrete structure itself.

Disaster resistance

Sun Block is fire rated at four hours, about twice the time of conventional 8-inch block. The block is reinforced like conventional block to withstand wind loads. Because it has slightly less load-bearing strength than conventional block, the manu-facturer sometimes recommends grouting some extra cavities to beef it up. Sun Block houses routinely meet Florida codes that call for windspeed ratings of up to 130 mph. The company is glad to look at your plans and advise you on this. If you're reinforcing anyway, the cost of doing more is minor—about three dollars per vertical cavity in a ranch house where grout is $40 per yard. Rebar, if it's used, would be another $1–$2. The block hasn't been used in a high seismic zone yet.

Product availability

Sparfil currently arranges for Sun Block to be produced at a plant in Tampa, Florida. All of the homes to date have been in Florida. Sparfil is willing to ship farther, but this increases the freight costs. In addition, the Sparfil representa-tives, who are also based in Tampa, would be less available to provide support.

Since there's only one plant and shipping times are longer to some places, it's necessary to order a little in advance. The builders farthest from Tampa found that the block they ordered would be delivered to them in 3–4 days.

Labor availability

There are no special labor needs for Sun Block. Regular masons use it without sig-nificant training or problems. The builders agreed that if you can put together the crews to build a conventional block house, you have the crews to use Sun Block.

Building department approval

Sparfil has put Sun Block through the standard tests for concrete block. If the building department has any question about it, the company's representatives will talk with them and give them copies of the results. Our builders reported that in their projects, the officials treated it like any other block and didn't re-quire anything special before giving approval, except sometimes the printed data from the manufacturer.

Required calendar time

The construction sequence in the projects we surveyed was exactly the same as it was for interior-insulated mortared block. The only exception was that some-times there was no insulation put between the furring strips.

Sometimes the block work went faster by a few hours because the block is about ⅓ lighter. A single mason could therefore lay more units in a day. In ad-dition, the cutting went a little faster.

Crew coordination

This was about the same as for interior-insulated mortared block. The one dif-ference is that electrical and plumbing subs found it easier to cut their own holes in the block.

Making connections

The builder-preferred fastener in Sun Block was an ordinary sheetrock screw. It goes in directly (no drilling required) and holds well. Nails also go in easily, but they can pull back out because the block material doesn't have the "grip" that lumber does. One builder claimed to get especially strong connections by gluing his wood (furring strips or 2-× lumber) to the block and holding it in place with drywall screws. Of course, plastic screw anchors and J-bolts work as well.

Joists for intermediate floors were generally hung on a ledger that was anchored into a bond beam with J-bolts. The roofing either attached to hurricane straps embedded in the bond beam or went on a top plate J-bolted into the beam.

Windows and doors could be attached to a buck, as with conventional blocks. But many builders went without a buck because it's easy to fasten directly to the block. Some builders used wood windows, screwing them to the front of the wall just as they would to frame. Doors could also be fastened to a buck or directly to the block. The decision of which door/window to pick and whether to use a buck depended partly on dimensions. If you use a door or window that isn't sized for the 8-inch module, to go without a buck you have to cut the block all around.

Utilities installation

Electrical and plumbing lines went in the same as they did into interior-insulated mortared block. Some pipes and vents were embedded in the foundation. The masons cut blocks and worked around them. Electrical and most small plumbing were routed through the furring strips after the wall was up.

When plans called for electrical lines in the block, this was easier than with conventional block. Electricians find that they can cut their own holes after the wall is up by pounding a good dent into the block with a hammer, then cutting it out with a utility knife. Then they can thread their cable (or conduit, if necessary) through.

Change flexibility

The builders we interviewed said that changes in the wall before it's reinforced are easier with Sun Block because cutting is faster. Changes that go through grout or rebar, or that involve filling in masonry, go about the same as with conventional block.

Manufacturer support

The builders we spoke with were pleased with the support they got from the Sparfil office in Tampa. There was little for anyone to learn to switch from conventional block to Sun Block. The Sparfil representatives provided what training was necessary on-site. They also helped with the plans and provided the local building departments with whatever they needed to approve the use of the block. Sparfil representatives were willing to come to the site if there were any problems, but this was rare.

Other considerations

The builders reported that their Sun Block jobs generated less waste because the block rarely breaks. When it's dropped on a slab, it simply dents. An 8-inch wall of the block achieved an STC of 48 in testing.

Further information

Sparfil Bloc Florida
P.O. Box 270336
Tampa, FL 33688
(813) 963-3794

6

Mortarless block systems

The mortarless block systems fulfill an old dream: to stack up the walls of a house as easily as a child stacks building blocks and then lock them together only after you've reached the roofline. The results are time savings, labor savings, less skilled labor, and the ability to make changes easily along the way. Yet these systems retain the advantages of concrete.

The mortarless block systems are very different from what used to be called *dry stack*. The old dry stack walls consisted of conventional blocks set on top of one another and then cemented together. The new mortarless systems use special blocks with multiple rows of cavities, not just one. This gives them flexibility. You can produce a basic, inexpensive wall, or you can fill cavities with insulation or reinforcing to get a range of R-values and levels of disaster resistance. That's why these systems can appeal to so many different market segments. Depending on what you put into the wall, its features and costs can meet the preferences of very different buyers.

IMSI

The IMSI system gets its name from the company that sells it, Insulated Masonry Systems, Inc. of Scottsdale, Arizona. Its basic 8-inch block has two rows of cavities with offset, cut-down webs. The cavities provide spaces for insulation, reinforcing, and utilities. There's also a 12-inch block with three cavity rows. The 12-inch block is used mostly in basements and commercial construction.

Figure 6-1 contains diagrams of all variations on the basic block. For 8-inch walls, the stretcher is about 75 percent of the units used. There are also half blocks (to go at the edges of openings), and there are right-hand and left-hand corner blocks. Each corner block has one large cavity that lines up vertically at the corner to make room for extra reinforcing. You can also line these blocks up at other points in the wall to house extra reinforcing or large utility lines. The 12-inch block comes in the same variations, except that only one corner block is necessary because the 12-inch block is symmetrical (one row of short cavities

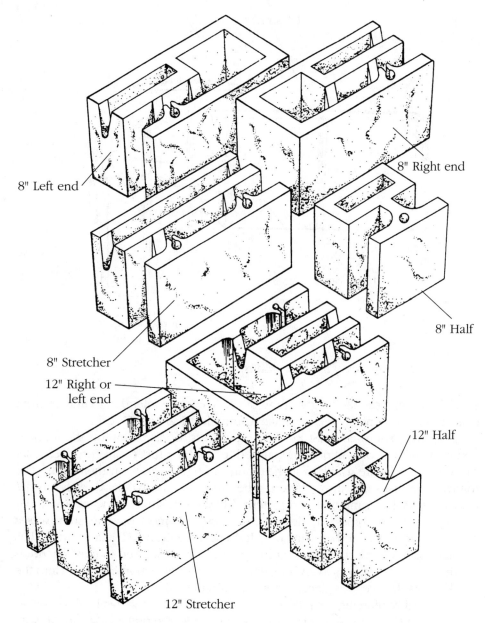

8" Left end

8" Right end

8" Half

8" Stretcher

12" Right or left end

12" Half

12" Stretcher

6-1 *Diagrams of the IMSI blocks.*

on either side of a row of long cavities). All blocks are full dimensions (8-x-8-x-16 and 12-x-8-x-16), not nominal.

In an 8-inch wall, the outer row of cavities is intended to hold foam inserts (pictured in Fig. 6-2) as insulation. The inner row is for conventional reinforcing and more inserts. Within the inside-most face of concrete are vertical, cylindri-

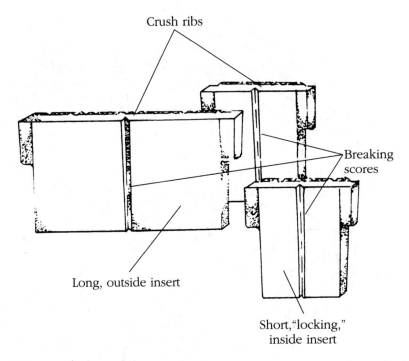

Crush ribs

Breaking
scores

Long, outside insert

Short, "locking,"
inside insert

6-2 *Diagrams of the IMSI inserts.*

cal channels that run the height of the block. These are intended to line up from
course to course and serve as a place to run electrical and TV cable.

Figure 6-3 shows an IMSI job site in progress. Construction off the foundation starts with a conventional mortar bed to level the first course precisely. Because the blocks are a full 16 inches long, no mortar goes on their ends. The
following courses are all mortarless. The crew sets the blocks directly on the
course below in a running bond pattern. They insert shims as necessary under
the corners of the blocks for leveling courses. The most popular shim was a
steel brick tie, pictured in Fig. 6-4. It's cheap, galvanized against rust, has
enough load-bearing strength to hold up, and can be stacked in twos or threes
if a thicker shim is necessary. After each course, the insulation inserts go into the
block just laid, except in places where reinforcing will go.

Standard reinforcing for IMSI is a 4-x-4 grid, plus one vertical cell on either
side of an opening and a lintel on top. Below grade, the builders we spoke with
used 2-x-4. The vertical reinforcing was every 2 feet, not 4. Reinforcing on the
4-x-4 grid is more than required with conventional block in many areas. But
that's what the system has most of its testing and code approvals for, and some
builders felt it added a margin of safety and sturdiness. The horizontal rebar cradles conveniently on the cut-down webs of the inner row of cavities.

After the walls are up, a plastering crew puts a coat of surface-bonding cement on each side, as shown in Fig. 6-5. This is a stucco-like mixture containing

6-3 *First few courses of an IMSI house. Rebar spacing is close because this portion of the wall will be below grade.*

fiberglass. You can buy IMSI's brand of cement or use Q-Bond, another brand. According to the manufacturer, just about any brand of surface-bonding cement would probably work as well, but because all the tests were done with these two, building officials could technically object to another one. The bonding cement is applied to a thickness of at least ⅛ inch.

IMSI is available throughout the western United States and Canada, and in parts of the East. The company is working to set up dealerships in the rest of the East in the near future.

6-4 *Brick ties placed beneath an IMSI block to level it.*

6-5 *Partially surface-bonded IMSI wall. The course at top is 12-inch block to form a brick ledge. Stacking will resume with 8-inch block above that.*

Because of the flexibility of its configuration, IMSI has sold successfully into several different market segments. It's possible to get an inexpensive, functional wall by adding only enough insulation to meet local requirements and adding no finishes over the surface cement. Configured this way, it has sold to some cost segment buyers, including government-sponsored affordable housing projects. When fully insulated and finished with fancier materials, it matches the preferences of quality buyers well, although there are some limitations on the architectural flexibility of the wall. Quality buyers have been the largest set of customers for IMSI homes to date. In some places, the houses have come in at about the same price as standard frame (even with full insulation and attractive finishes) and have gotten value market buyers. Its fire resistance and the provisions for reinforcing also allow construction of a highly disaster resistant house at relatively little extra cost.

Labor and materials' costs

The IMSI builders in the West and the Plains States we talked with pretty consistently said that they were bidding about the same or slightly less than other builders offering 2-x-6 construction with a wood siding exterior and sheet rock interior. The IMSI walls in these comparisons were usually sided with stucco on the exterior and a direct coat of joint compound or plaster inside. A few builders even told us that they had quoted almost even with 2-x-4 frame. But when the 2-x-4 was sided with vinyl, the IMSI generally came in higher. One quote was 10 percent more for the complete wall, or around $1500–$2000 more for a typical 2000-square-foot (living space) home.

Table 6-1 contains some typical unit cost quotes. If you were constructing the bare minimum functional wall, you could omit all finish work (exterior finish coat and painting, interior finish and painting). By the figures in the table, the remaining cost would be $4.10 in an area requiring no insulation, $4.95 where the outer insulation only was required, and $5.75 with full insulation.

In more fully loaded versions of the wall, quotes for the total cost usually ranged from $7.00 to $8.00 per square foot of area, including paint inside and out and everything in between. A lot of the variation came from differences in local labor rates. But shipping costs can also vary widely. The insulation inserts and surface bonding cement are currently manufactured in Phoenix and Colorado Springs, Colorado. IMSI ships them from those locations. The local IMSI dealer usually arranges with a local plant to run the block, but in rare cases it's also shipped in at added expense. The entire package of IMSI materials has sold in the Colorado Springs area (where there are minimal shipping charges) for about $2.75 per block, which translates into $3.09 per square foot. For a house built in Bismarck, North Dakota before local block manufacturing was set up, the price was $3.75 per block, or $4.22 per square foot. The figures in Table 6-1 are for a job with nearby block manufacturing, but inserts and cement shipped about a thousand miles.

Table 6-1.
Representative Costs for Exterior Walls of IMSI Block[1]

Item		Cost per Square Foot
Blocks		$1.25
Insulation inserts		0–1.65
None	$ 0	
Outer cavities only	.85	
Outer and inner cavities	1.65	
Surface bonding cement		.40
Grout and rebar		.15
Stacking and reinforcing labor		1.15
Surface cementing labor		1.15
Subtotal, structure and insulation		$4.10–5.75
Exterior finish		$0-1.20
None	$ 0	
Paint only	.50	
Stucco and paint	1.20	
Interior finish		$0-1.20
None	$ 0	
Paint only	.80	
Texture coat and paint	.90	
Smooth plaster and paint	1.20	
Total, including finishes		$4.10–8.15

[1] The costs listed here are in the middle of the range of quotes we received from builders reporting their actual project costs.

Learning costs

The builders all agreed that they and their crews felt efficient and comfortable with the system after two to four houses. One said that his labor cost for stacking and grouting was $1.75 per block on his first house, but has come down to $1.00 since then.

They said the things that they and their block crews had to learn were mostly correct leveling and reinforcing. Leveling the block isn't hard, but if you don't do it precisely on every course, the errors accumulate and you have to waste a lot of time unstacking and redoing your past work. It starts with a precise footing. The builders said concrete crews can lay these if they want to, but sometimes they don't appreciate how exact they need to be. We heard one story about a contractor who insisted he could maintain level within ¼ of an inch all around (about what you need) with a trench footing. He ended up being off as much as ¾ of an inch at points. The block crew had to do a lot of shimming that took time and cost hundreds of dollars in labor. The lesson the builder took away was always start with a formed footing. Above the footing, the big lesson was to use shims as necessary to level every block. They're cheap, the work's fast, and the consequences of being out of level are expensive.

Placing the reinforcing steel correctly takes some getting used to. This is mainly because the reinforcing alongside openings is usually one-half block away from the opening, instead of directly alongside. When no buck is used, the half-cavities right along the opening are open to it, so grout could flow out of them.

From the supervisor's point of view, one other thing to get used to is the speed at which the block goes up. "It's so fast," said one builder, "if you're not on top of things, you can go by some point where you're supposed to put in rebar or inserts and totally miss it."

The electricians were the other group that had some learning to do. Because the walls are almost never furred out, they're putting a lot of cable and outlets in the block. Our builders generally talked new electricians into charging at the same rates they charge with frame. But the electricians reported only breaking even on the first house, then getting as efficient and profitable as with frame on the third or fourth house.

The manufacturer and its dealers help builders learn the system. One of them will send a trainer to the job site for the first day of block work to help out and give instructions.

Buyer reaction

IMSI dealers and builders have had a lot of success pitching the system as a superior building product. They present it as faster, stronger, maintenance-free, a third more energy-efficient, and with lower insurance costs than frame. In some areas the insurance bill is, we hear, about $200 a year less for a typical 2000-square-foot house because of lower fire and wind premiums. Most buyers are in the quality segment. However, some were clearly value segment buyers, and a few became interested because of the system's resistance to disaster.

Buyers in the north appear to accept the stucco finish of the IMSI block more readily than we have found them to accept it with some other systems. We have no good explanation for this. Perhaps they have focused more on the functional benefits and less on the aesthetics. In one area the unusual architecture of the first few IMSI houses built actually got attention and inquiries about the block.

Exterior and interior finishes

Ninety percent of the time. the exterior finish is a final coat of the surface bonding cement. This final coat has no fiberglass in it. Otherwise, it has the same ingredients as the structural coat beneath it. It's part of the standard package of materials the manufacturer offers, and it's included in the cost figures of Table 6-1.

The finish coat can be textured like ordinary stucco. Some builders formed details and popouts by gluing pieces of foam to the outside after the structural coat was up. The finish coat went over the foam just as it did over the rest of the wall.

The finish material comes in white and gray. If one of these colors is acceptable, no more coloring or painting is necessary. Otherwise you can paint. Builders we talked with said ordinary exterior latex works fine.

You can also leave the exterior with only the structural coat. No finish coat or paint are required. We are told the joints between the blocks might show through as slightly dark lines. But you'll have a fully functional white (or gray) wall.

Occasionally a builder added a brick veneer to an IMSI house at about the same cost as a veneer on frame. One house we heard of used furring strips on the outside to attach clapboard. The cost of this should be the same as on conventional block—about $.20 per square foot for the furring, plus the cost of the clapboard. With any of these extra sidings, you can leave the finish coat and paint off the IMSI wall.

Inside, most builders opted for a smooth coat of plaster or joint compound, or a sprayed-on texture coat. These are both inexpensive to add because there's no wall board or extra surface preparation over the interior structural coat. And if your block and surface-bonding workers are reasonably skilled, the surface will be even. It's possible to omit the plastering, painting, or both. Without plaster, the inside wall will be a little rough, but, of course, cheaper.

Design flexibility

When the house designs had basements, our builders made them out of IMSI, too. According to the manufacturer, if you do this you don't need waterproofing because the exterior structural coat keeps water out, but most builders add conventional dampproofing anyway because it's cheap and they want to be safe. A lot of the builders were quick to point out that with IMSI you get a finished basement at almost no extra cost. It's waterproof and has an acceptable finish with only the structural coat. If you want, you can add insulation in one or two rows of cavities and plaster the inside at a low extra cost. The result was that they could offer an excellent finished basement for a few extra dollars, and they pitched it successfully to customers.

Houses built of IMSI have gone as high as a basement plus three stories. The manufacturer claims you could go higher with proper reinforcing. The costs probably don't go up quite as much for extra floors as they do with conventional masonry because there's no need to shovel mortar up.

Curves and unusual angles can be handled as with conventional block. Gentle curves are formed out of half blocks stacked in a stack bond, each one angled slightly. The surface cement must then be put into the gaps between columns on the outside. The curve also leaves gaps in the insulation inserts. You can stuff in bits of foam to fill these. Masons can form odd angles by cutting the block before stacking. You then have to do some ad hoc cutting and stuffing to get the EPS inserts in. Sometimes builders handled curves and angles by simply framing those portions of the wall. That's what they almost always did with odd-

Interior of a cavity-insulated home built with colored half-high block.

Interior-insulated house of colored, half-high, split-face block under construction.

Interior Insulated block house in upstate New York. Combines smooth and split-face block.

A home of Superior Walls finished with paint only.

A finished home in Arizona built with the Integra Wall System.

Florida house built of Intralock Block.

A building of GREENBLOCK sided with stucco.

Entry to a Florida house built of Royall Wall Systems panels, including the columns and arches of the entryway.

A winter retreat built of Sparfil Wall System II.

A Florida home built of Sun Block.

Home constructed of Hebel block.

Insulated Masonry Systems, Inc. (IMSI) house in Utah.

Stone veneer house built with Lite-Form.

Completed Georgian house built of 3-D panels. The curved half-walls along the entrance are also built of 3-D.

Home built with 3-10 Polysteel.

Completed exterior-insulated home.

shaped openings. They framed the odd opening inside a larger square one formed out of the block, covered the frame with metal lath, then put the structural and finish coats over everything.

R-value and energy efficiency

The 8-inch block measured R-13.6 in a guarded-hot-box test. The test was done with a fully insulated, unreinforced wall, so it took into account the heat loss through the webs. There would be some additional losses through reinforced sections of the wall, including the 4-x-4 grid and reinforcing along either side of an opening and in the lintel. However, because the reinforcing normally is on the inside row of cavities only, there's still the insulation of the outer row at these points. Places that might have no insulation are the large end cavities of the corner blocks (normally filled solid with grout and rebar) and columns of corner block inserted in the middle of a wall for extra reinforcing or running large utility lines.

The company has had an independent engineering firm calculate mass-corrected R-values that use the guarded-hot-box value of 13.6 as a starting point. The results ranged from a value of 23 for central Florida to 15.5 for central North Dakota.

The manufacturer has measured the block's thermal mass at 6.18 Btu/sf/degree Fahrenheit for the 8-inch block, and 8.85 for the 12-inch. With these numbers, the wall qualifies for special R-value treatment by most energy codes.

An IMSI house owner in Arkansas says that his house is quite similar in size and shape to his neighbor's conventional frame house, and they both have similar HVAC systems and lifestyles. Yet, he claims, his meter reader tells him that his gas bill is consistently one-third lower. A couple in the Provo, Utah area have a 4000-square-foot IMSI house. They heat the entire home at a consistent 76 degrees during the day, then shut off the furnace and open their bedroom windows at night. They say that in the dead of winter the indoor temperature only falls 5 degrees by the morning. The highest heating bill they've had for any one month was $95.

Thermal mass

Because there are two insulation envelopes, exactly how much of the thermal mass is inside the insulation is a matter of judgment. If you count only the concrete inside both layers of insulation, it's about 2 Btu/sf/degree. If you count the concrete between the two layers as well, it's about 4 Btu/sf/degree. The builders we talked with had not sold houses to customers using them for passive solar or load-leveling applications, so we don't have any performance stories to report on these uses.

Water resistance

Insulated Masonry Systems claims that the continuous surface bonding cement on the outside alone provides good resistance to moisture penetration. In the

1970s, manufacturers had labs perform the wind-driven rain test on dry stack walls using similar cements. The results showed only insignificant amounts of water getting through the wall. The addition of a finish coat and paint to the outside should add even more water resistance. We heard no reports of water penetration from IMSI homeowners. The resistance of the surface cement to cracking (discussed in the next section) might contribute to keeping it a good water barrier.

Maintenance required

IMSI appears to be about as close to maintenance-free as any system we encountered. Both the outside and inside are backed everywhere with solid concrete, so damage from impact is rare. Builders and homeowners alike claimed that with the fiberglass-reinforced surface cement, there's virtually no hairline cracking.

The one warning we got was that the surface bonding has to be correctly applied to get these results. That means that, like other cement products, it has to be kept moist for a while. According to the manufacturer's construction manual, wetting it down a couple of times in the 24 hours after it's up is sufficient, except that in dry climates it should be misted twice a day for 48 hours. If this isn't done, cracks and weak spots can develop.

Some builders claim that ordinary exterior latex paint "bonds" with the finish coat material. It might therefore never peel, although some colors might fade in the sun. But it's hard to verify this because the oldest IMSI buildings are now only nine years old. It's probably safe to say that painting won't be necessary any more often than over other concrete and stucco surfaces (five to seven years).

Disaster resistance

An unreinforced 8-inch wall with a heavy load on it survived the fire test for 2 hours and 20 minutes. At that time it developed a crack. The manufacturer expects a reinforced wall to go longer but hasn't had one tested yet. Because they're almost always reinforced, the performance of a reinforced wall would probably also be a better indicator of fire resistance for an IMSI house.

The 4-x-4 reinforcing normally used in the system is more than is normally required by the nation's strictest wind codes. IMSI houses have been built in southern Florida, which has arguably the strictest codes of all, with the standard reinforcing.

The 4-x-4 grid also qualifies in most cases for earthquake Zone 4. IMSI was used to construct the new U.S. Army armory in Stockton, California. The building had walls over 20 feet high. It was less than 30 miles from the epicenter of the 1989 earthquake (the one that interrupted the Oakland A's-San Francisco Giants World Series), which measured Richter 7.1. Although the manufacturer was not allowed to see the building afterward because it's a military facility, Army officers sent a letter saying that there was no damage to the structure.

Product availability

There are currently dealers for the system that cover the West from the Plains States to the Pacific Coast. In addition, there are established dealers covering Arkansas and Kentucky, and new ones about to begin operations in Florida and parts of Canada. Insulated Masonry Systems is working to sign up others to cover the entire country in the near future.

Once dealers get established, they almost always have local plants lined up to make the block. Sometimes they have to ship it in for their first few orders. The inserts and surface-bonding cement and finish coat are shipped by Insulated Masonry Systems from Phoenix and Colorado Springs. The company is planning to add a third manufacturing facility in Kansas. In addition, you can buy Q Bond or some other brand of cement/finish if you prefer.

Labor availability

The builders found they could get the stacking done with one or two skilled supervisors overseeing three to five laborers. The IMSI dealers sometimes sent out experienced crews to do the stacking and reinforcing for builders who preferred to have someone else take care of it for them. Otherwise the builder picks his own crew. It's usually a masonry crew, but in some cases they've used carpenters. The dealers or Insulated Masonry Systems will send out someone to train the skilled supervisors and sometimes stay around for the first day or so of work. Note that if you want to use your own crew, it has to go through this training. IMSI dealers won't sell you the system otherwise.

Different crews take to the system differently. The key thing to look for seems to be enthusiasm. When builders used a carpentry crew, it was because they were interested and motivated, and they picked it up well. On the other hand, we got no reports of any crews that simply could not do an adequate job. Maybe the builders screened out the crews that were really opposed before the job started.

This system also requires capable plasterers. This is true regardless of what finishes you use because they have to do the surface bonding. But our builders always managed to find a crew by just asking around, and the plasterers didn't have much to learn.

Building department approval

IMSI now has a complete ICBO evaluation report (ICBO Report No. 4997). The builders say this is enough to get quick approval in the West, and usually in the East as well. It looks pretty much like reinforced masonry to most building departments, and that's how ICBO classifies it. In a few cases an eastern building department has asked for an engineer's stamp. In those cases the local dealer has had a local engineer review and stamp the plans for the builder.

The building inspector might want to see the reinforcing steel before each bond beam is poured. In that case you have to figure into your scheduling one inspection every 4 feet. This can be tricky because the wall goes up fast.

Required calendar time

Inexperienced crews of five have stacked and reinforced all the walls of a large house (2200 square feet of living space) in six days. Builders generally report that, with an experienced crew, a job this size drops to about four days. The stacking can proceed in almost any weather, and roofing can go on as soon as the blocks are up. Some builders prefer to put the floors and roofs on before the surface bonding to compress the walls fully.

The surface bonding on an average-sized house takes a two-person crew about two days inside, plus three to four days outside. The outside takes longer because there are often ornamental details and there are usually two coats. Like other field-applied concrete, this has to be done in weather that's above freezing, or maybe a little below if special measures are taken.

Crew coordination

How many crews are required depends a lot on who the builder picks to do the work. Some builders used one crew to lay the basement out of IMSI block, lay the walls above-grade, and do the rough carpentry. And some had the same crew do interior plastering and the surface bonding and exterior finish. Depending on how you usually work, these practices could reduce your total number of crews by as much as two. Or, if you decide to have all of these things done separately, you might end up with one or two extra crews.

The one coordination task that former frame builders found they had to pay special attention to was the same one with conventional block—providing for the electrical lines. A couple of builders handled this by personally marking the location of boxes on the second course of block with a crayon. According to one of them, this takes "about 15 minutes" for the entire house. The stacking crew would cut the blocks and insert conduit at each mark on the third course.

Making connections

Floors are connected as with conventional masonry—with pockets or J-bolts in the bond beam that hold a ledger to the inside of the wall. If long J-bolts are used, they have to anchor into both the inside and outside cavities. Or they can anchor into the large cavity of corner block that's been set in the spot. For roofs, the choice was usually to put J-bolts in the top bond beam and screw a top plate to them.

The lintel above a window can be steel. It must have a vertical rib that's off-center. Some brands offer this as a standard product. The blocks that rest on the lintel must be turned upside down so that the cut-down end of the webs face the lintel. These two steps allow the rib of the lintel to protrude into the outer cavities.

But most builders preferred to form the lintel out of the block, as pictured in Fig. 6-6. They erected a temporary header supported by 2-x-4 braces or a permanent buck to hold the block. They put rebar in the inner cavities of the block and turned it upside down before placing it on the header or buck. After grout-

6-6 *Window opening in an IMSI wall with a lintel grouted in place. If the builder had wanted to, he could have slid the flanges of the window down the continuous channel of foam inserts along the jambs of the opening.*

ing, the lintel was firmly in place, and the surface bonding strengthened it further. They preferred this method because it didn't have a steel lintel to leak heat out (or in), and in some cases because it made it possible to slip the flange of the window into the cavities (described in the following).

For windows, the most popular attachment method was to slide the window flanges down the outer cavities of the block on each side of the opening before the lintel was on. You can just wedge the flanges in between the outer face shell of concrete and the inserts in the cavities. The bottom flange slides into the cavities across the bottom of the opening, and (if you make your lintel out of upside-down block) the top flange fits inside the cavities above. Some builders felt that having foam hold the window in place was not secure. They cut away the inserts immediately around the opening and replaced them with 2-x-3 lumber. If the lumber were put into the inner cavities, the crew could put J-bolts or some other fastener through the 2-x-3. Then during grouting they would grout right up to the opening, gripping the fasteners and firmly anchoring the lumber. The window would slide between the wood and the outer concrete face.

Windows could also be attached with standard bucks. Or they could be attached like standard wood windows. The window flanges would overlap the opening and get fixed to the wall from the outside with masonry screws.

Doors could be attached to a buck. Or they could be fastened through the frame into the concrete or 2-×-3s set into the cavities around the opening. Both methods could also be used in combination: the buck might nail into 2-×-3s set into the cavities.

Cabinets and interior fixtures were attached with standard concrete fasteners. Favorites were plastic screw anchors and concrete screws. You have to drill in concrete, but you don't have to hunt for a stud.

Utilities installation

Electrical wiring can go in the preformed channels or in standard conduit. Figure 6-7 shows a detail of boxes mounted for cable that runs through the channels. The wall crew cuts a hole for the box centered along one of the channels. They can put the cable through as they build, or the electrician can fish it through later. But they have to make sure the channels are precisely aligned from block to block and no debris gets inside. They can run a rebar down a channel to clear it if necessary.

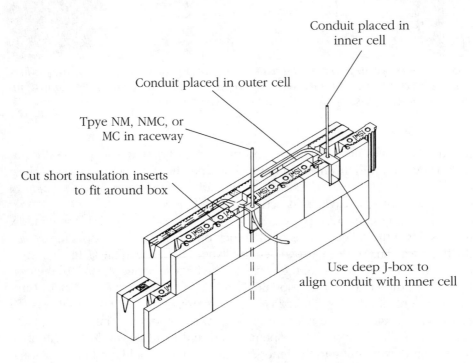

6-7 *Detail of electrical box placement that uses the preformed channels in the IMSI block for the cable.*

But most builders we talked to ran standard conduit through the inside cavities instead. Some codes don't allow bare cable in block. Using conduit is done in the same way as it is with conventional block.

Change flexibility

Builders pointed out that they can rearrange the wall easily up until the time they grout it. While you're waiting for the ready-mix truck, there's a good opportunity to double-check the plans and the plumb and level of the walls. Moving things around at that point is even easier, faster, and cleaner than it is with frame.

After grouting, but before surface bonding, it's still fairly easy to move things within one of the reinforcing squares of the 4-x-4 grid. You can prop blocks up above where you're moving and push out or slide in blocks as necessary.

Making changes across a grouted cell presents the same difficulties as it does with other masonry. One IMSI builder told us about moving an opening that had been misplaced 16 inches to one side. This required cutting through the reinforcing alongside. Two workers made cuts in the block with a diamond blade saw and knocked it off the grout with a sledgehammer. This left the grout and rebar standing. They knocked the grout off and cut the rebar with a portable band saw with a hardened blade. It took them three hours.

Filling in block is relatively easy because there's no mortar. But you might have to replace the reinforcing somewhere else in the wall, which is more difficult because you have to find ways to get the grout into the middle of a standing wall.

The builders we talked with rarely had to make changes after reinforcing. They simply learned to plan their layouts carefully and make adjustments before the pour.

Manufacturer support

Help is available from Insulated Masonry Systems in Scottsdale and from the regional or local dealer. They provide written manuals on how to use the system that builders say are clear and well worth reading. One builder said, "I memorized the information. It was very helpful. That's where I learned almost everything."

The company or a dealer will also send a representative to train builders on their first house. This can be up to two days of on-site instruction. It's free of charge. All agreed it was helpful, but one builder said, "It's all new, and when they train you it comes at you so fast. You really have to build one or two to understand completely what they're telling you. And then I came up with my own tricks that even they didn't teach me."

Other considerations

The IMSI wall tested at a very high sound transmission class (STC) of 61. Since the system is mortarless, there are no mortar drippings around the site. There are block cuttings, as with all masonry. There are also loose pieces of cut inserts, but these are light and easily collected.

Further information

Insulated Masonry Systems, Inc.
7234 East Shoeman Lane, Suite 1
Scottsdale, AZ 85251
(602)-970-0711

Intralock

The Intralock block system is bound together in a unique way. Instead of reinforcing at 4- or 8-foot intervals and then surface bonding continuously over front and back, the block contains an inner row of cavities that is poured solid with grout and rebar to lock every unit into place. In addition, the blocks are precisely sized to true 8-x-16 dimensions for quick stacking. The result is an extremely strong wall that erects faster than almost any other system. The amount of labor is so small that the total cost of the wall is about dead even with conventional block. It can be lower than frame in areas where there are strong wind requirements. Those drive up the cost of frame, but not of Intralock.

The Intralock system is the invention of the Intralock Corporation in Pompano Beach, Florida. Currently the blocks are manufactured only at the company's plant there. However, the company has licensed a company in Hawaii to produce them there, and is negotiating with other block producers to offer the system in their parts of the country, too.

To date, most of the buyers who became interested in homes of Intralock were attracted to the general quality of construction or the prospect that the especially strong wall would resist wind well.

The blocks of the system are pictured in Fig. 6-8. The most-used is the stretcher, shown again in Fig. 6-9. It has three rows of cavities. The middle row is filled with grout after the wall is stacked. It's designed so vertical and horizontal re-

a   **b**

6-8 *The Intralock blocks: (a) half block, (b) conduit block, (c) bond beam block, (d) corner block, (e) stretcher, (f) bond beam corner block.*

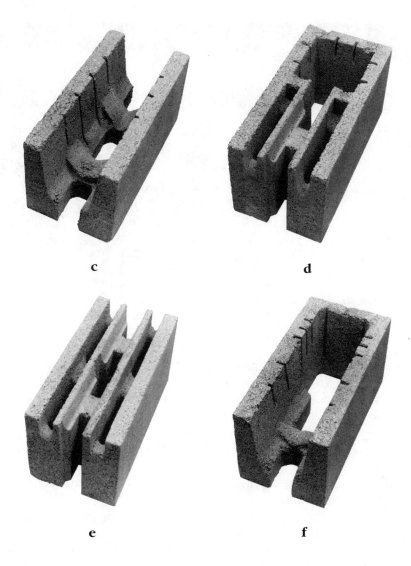

c

d

e

f

6-8 *Continued.*

bar placed into it are held in place without any tie wires. The bars can go in as close together as every 8 inches. The two rows of cavities on either side of the center are identical. They can remain empty for air circulation through the block, hold electrical and small plumbing lines, or take insulation. There's also a half block (with a similar set of three cavities) for wall ends at openings, a conduit block that's hollow to hold large utility lines, and a corner block with an end that's open like the conduit block to hold heavy reinforcing. The corner block is called "omnidirectional" because it can be used on either side of a corner. To open up a channel for grout to flow, you knock out a chunk of the block on the side that faces the connecting wall. There are convenient scores on both sides to make it easy to take out a piece with a hammer. The conduit block has similar scores.

6-9 *Intralock stretcher block stacked at the bottom of a window opening. The center row of cavities can hold vertical and horizontal rebar and will be fully grouted.*

The manufacturer has recently added a "bond beam" stretcher block (pictured again in Fig. 6-10) and "omnidirectional bond-beam corner block" (Fig. 6-11). These have one row of wide cavities that narrows down at the bottom of the block to match the middle row of cavities in the stretcher. They're designed to be the top course of block. You pump the grout into the wide opening of the bond beam block, the grout is channeled into the center row of cavities in the block below, and the bond beam is grouted solid. The bond beam block also has special slots on either side into which you can slide metal cradles to hold the bond beam rebar in place.

Figure 6-12 pictures a job in progress. The first course of Intralock block is mortared conventionally to the foundation to level it precisely. After that, the crew stacks block. Some typical sections are shown in Figs. 6-13 and 6-14. The manufacturer grinds the block to dimensions so precise that the builders we talked with swore they virtually never used a shim. One went up to 22 feet on a commercial building, and the builder claims that the top of the wall was plumb and level without shims.

Some sort of brace is necessary to hold up the block above an opening before the grout pour. Some builders used temporary lintels made of 2-x-8s held in place with 2-x braces or screw jacks, as in Fig. 6-15. One built frames to the exact window or door size, set them in place when the wall was up to the bottom of the opening, and left them there as a guide for the stackers to work around. The block above an opening must contain rebar to form a lintel. It can be ordinary stretcher block or, if more strength is necessary, bond beam block

6-10 *Intralock bond beam stretcher block. A metal brace has been placed in preformed grooves to serve as a cradle for horizontal rebar.*

6-11 *Intralock omnidirectional bond beam corner block. The shell has been knocked out along the scores on one side to allow grout to flow around the corner.*

6-12 *Intralock house in progress.*

6-13 *Partially stacked Intralock wall. The corner block at the top has one end open to hold extra reinforcing, and will have part of its side knocked out along score lines to allow grout to flow around the corner.*

6-14 *An Intralock corner block used in the middle of a wall to provide space for extra reinforcing.*

6-15 *Temporary lintel using lumber braces and a screw jack to hold up the Intralock block above.*

with multiple bars in it. Because all the block fill with grout in the pour after the wall is up, there's no need to do an early grout pour to fill a lintel that's in the middle of the wall.

When the wall is completely up, grout gets pumped in at the top. You must use grout that contains only fine sand aggregate and includes a super-plasticizer additive. This formula is meant to guarantee a very thin mix that flows throughout the fairly narrow center cavities and fills them completely. Any good concrete vendor should be able to give you a suitable mix if you explain the situation and ask for grout with an 11-inch slump. The people at Intralock Corp. can advise you if you're having trouble explaining it. The grout costs about $52 per cubic yard in South Florida, versus $38 for conventional grout.

Labor and materials' costs

The South Florida builders who have used the system agreed that it costs the same or slightly less than conventional reinforced block. The materials are more, but the labor is less because the wall goes up faster and you can use workers who are less skilled. Typical unit costs we got from the South Florida custom builders we interviewed are listed in Table 6-2.

Table 6-2. Representative Costs for Exterior Walls of Intralock[1]

Item		Cost per Square Foot
Blocks		$1.25
Grout and rebar		.45
Stacking and reinforcing labor		1.30
Furring strips and ¾" insulation		.30
Subtotal, structure and insulation		$3.30
Exterior finish, materials and labor		$ .80–1.10
Elastomeric paint only	.80	
Stucco and paint	$1.10	
Interior sheetrock		.80
Interior painting		.70
Total, including finishes		$5.60–5.90

[1] The costs listed here are in the middle of the range of quotes we received from builders reporting their actual project costs.

The 8-x-16 blocks (stretcher, corner, and bond beam) cost $1.05 per piece, compared with about $.60 for a conventional unit in South Florida. The 8-x-16 units (half and conduit block) cost $.75. The builders we spoke with used stucco on the exterior. The insulation between ¾-inch furring strips is standard for South Florida. Adding more would obviously add expense.

For a very inexpensive but functional exterior, the manufacturer claims you could get away with a coat of elastomeric paint alone. The block joints might show through, but the paint should seal the wall effectively. We heard some speculation that even an unpainted wall might be adequate, although no one so far as we know has left one unfinished. Inside, the norm is to use furring and sheetrock, although you could probably get by with ordinary paint alone if you didn't need insulation. From our figures, a bare-bones wall with elastomeric paint outside and latex inside would cost $4.80 per square foot.

Learning costs

Our builders agreed that learning the system is very fast. One hired a masonry crew and found that, "They were slower than with regular block on the first day, but after that they flew. With regular block they can do 250 a day per mason, but with Intralock they can stack 500 easily after the first day." Another who put together his own crew of unskilled laborers claims that they were up to full speed immediately. He learned the system well in advance, and the laborers simply did what they were told, so there wasn't much learning to do on the job. Bear in mind, however, that these builders were in South Florida, where most workers are very familiar with block construction. Converting workers from frame to Intralock might involve some more start-up learning.

Intralock gives first-time users a three-hour training session in the use of the block. These are held at the company's yard in Pompano Beach. There's a slab set up to stack a small house. According to the president, "We show them how to stack it up, then we tear it down again to show the next group." Intralock Corp. will only sell the system to trained contractors.

Buyer reaction

The buyers of South Florida generally prefer concrete block construction over frame, and so they react well to Intralock. The ones who chose Intralock were usually impressed with the opinions of their contractors, engineers, and architects that the wall would be exceptionally strong. But in a few cases, what made the difference was a promise from the builder that the house would be up much more quickly than conventional block. Speed was very important to these buyers. (And the houses did go up faster.)

The look of the Intralock wall had no bearing on the buyers' decisions. The houses were all stuccoed outside and sheetrocked inside, just like 90 percent of the houses in southern Florida.

Exterior and interior finishes

All houses we are aware of were stuccoed outside and furred and sheetrocked inside. According to the manufacturer, an Intralock wall can be veneered with brick or other masonry, or furred to attach vinyl, aluminum, or wood sidings just as readily as is done with conventional block.

Design flexibility

Intralock has never been used for basements (mainly because there aren't many in Florida), although the manufacturer believes it could be.

It's been stacked in commercial buildings three stories high. The cost probably goes up with extra stories, as it does with conventional masonry, except that there's no need to shovel up the mortar.

For odd angles, one builder simply cut the block as you would with conventional masonry. This required some precision to line up the center (grout) cavities. Another builder with curves on his floor plan framed that section of the wall instead. Intralock Corp. has started to deliver a 45-degree version of the block. The odd-shaped openings (arches, nonrectangular windows) were always framed into a larger, rectangular opening in the block.

R-value and energy efficiency

All builders we interviewed put an R-3 batt insulation between their interior furring strips. One also added vermiculite into the outside row of cavities. It's also possible to add foam or loose fill insulation into the inside row, although you have to take into account any cuts you'll make in the wall for utilities. If you do add insulation to the cavities, there will still be a few breaks in it: the two webs in the block on each cavity; any parts of the wall reinforced solid from front to back (including the bond beam, corners, and any lintels formed out of the bond beam block); and columns formed of conduit block.

The manufacturer has not yet tested the R-value of the wall, with or without insulation. It's currently working to calculate an R-value for the block alone. Because the Intralock block has more concrete than a conventional block, it also has more thermal mass. The weight of the concrete is about 53 pounds per square foot of wall before grouting, and more than 60 pounds after. Using standard conversion numbers for the thermal mass of concrete, we estimate the block's thermal mass at more than 11 Btu/sf/degree Fahrenheit per square foot. With grout, we estimate it at a little more than 13. This is far in excess of the 6 needed to qualify for reduced insulation requirements under most energy codes.

Thermal mass

As previously noted, an Intralock wall has a large thermal mass. If insulation were installed between furring strips inside, all of the thermal mass would be outside the insulation envelope. If instead insulation were placed into the inner row of cavities, about a quarter of the mass would be inside the insulation. If insulation were placed in the outer row, about three quarters of the mass would be inside. We did not encounter any Intralock houses designed for passive solar or load leveling effects.

Water resistance

The builders and occupants we spoke with complained of no problems of water penetrating their walls. All houses had the protection of a coat of stucco plus a coat of paint. In addition, the manufacturer claims some additional water-protection features of the block. There are two rows of empty cavities (if not filled with insulation) through which any water that penetrates can drain downward. And Intralock will put waterproofing agents into the block if requested. It's also possible to put waterproofing agents into the grout, and we're told that some builders have done this, too.

Maintenance required

The builders and buyers we spoke with claimed there was never a crack in the stucco. They thought this was probably because the wall was so solid it didn't shift like frame or block that's grouted only intermittently. This left periodic repainting as the only maintenance expense.

Disaster resistance

Independent engineers have calculated the fire rating of the Intralock wall at four hours. A commercial building made of Intralock did experience a major fire inside. It housed a furniture store, and the flames were fueled by large amounts of rattan and plastic. The fire melted the steel roof trusses. But the wall and its stucco exterior survived intact. The only repairs required on the wall were a cleaning and repainting.

The system exceeds the wind requirements of Dade County, Florida, and these are some of the most stringent in the country. It has not been tested or examined by engineers specifically for earthquake resistance, nor have any Intralock buildings been constructed in high seismic areas.

The incremental cost of adding reinforcing to satisfy almost any level of wind or earthquake requirements is small. Because the center cavities are grouted solid anyway, the only extra cost is the additional rebar (about $3 for an 8-foot #5 bar) and the labor to place it as the wall goes up.

Product availability

Intralock blocks are currently available from the Intralock Corporation plant in Pompano Beach, Florida. To date, all buildings constructed of the system in the United States are within 200 miles of there. Shipping costs would mount for farther building sites. The manufacturer is in discussions with other block producers in various parts of the country about licensing the right to make and sell the blocks. As of press time, an agreement was in place for a plant in Hawaii to produce them. All other parts of the system are generic products that are widely available: rebar, grout, superplasticizer, and so on.

Labor availability

The builders we spoke with found that good masons took well to the system. The blocks are heavier than normal (about 47 pounds, versus 30–35 or so for conventional units), but the masons don't need a trowel, so they can lift with two hands. The builders found that they could get the masons to accept the block and work at a good clip by giving them incentives on the first couple of buildings. These incentives were all based on the faster rate at which masons can lay Intralock block. One builder told the masons they could quit each day as soon as they laid so many blocks. Because the masons wanted to leave, they worked quickly, and they laid the block right because they didn't want to have to lay it again at no extra pay. Another builder let the masons mark up the Intralock more. Between that and the greater speed of laying block, they pocketed a lot more money in the same amount of time. Even paying the same per block will give the masons more money on the fast-stacking Intralock. Pointing that out seems to help.

One builder had good success hiring nearly all unskilled labor and supervising them himself. Only the builder and one reliable helper had to know the details of the system. The laborers picked up what they had to know as they went along.

The manufacturer says that carpenters also take well to the system. We didn't speak with carpenters that laid it, but we did notice that the part of the operation that's the hardest for nonmasons (mortaring) is not necessary in this system.

Building department approval

The Intralock blocks have an evaluation report from the SBCCI (No. 83205), which writes the "Southern" or Standard Building Code. No special steps were required for building department approval in any of the houses we asked about.

As with conventional block, the building inspector might demand to see that the rebar is all in place before the grout pour. However, the inspector does not normally require that there be lookouts cut in the block on the bottom course. If the builder shows that the grout used will be very thin (11-inch slump), it's a safe bet that it will flow to the bottom without voids.

Required calendar time

One thing the builders agreed on is that Intralock is fast. They all got their exterior walls up, from slab to roofline, in a day. This included rebar and all openings, but not grout. One used a ten-man crew of nearly all laborers to stack the walls to a ranch home of 3200 square feet of living space in eight hours.

The grout takes another couple of hours, including setup. It's usually scheduled for the next morning. After grout, the rest of the work can proceed almost immediately. The roof can go up because the wall can take a load even before the grout hardens. The only jobs to hold off on are the ones that could put a severe side load on the wall (such as backfilling) before the grout has a couple of days to set.

The builders especially liked getting the structure up in one day. They didn't have to worry about delays that could arise because the crew had other jobs to do. There was no chance the crew would complete half the wall, then not show up for three days.

Stacking can proceed in any weather. It's hard for the crew in heavy rain, but it can be done. Extremes of heat or cold have little effect on the system, so long as the workers can stand the conditions. The manufacturer claims that the grout pour can also occur in wet block because the grout tends to push any water that's inside the block out through the joints. In cold or dry weather, the usual measures used for grout work on conventional block apply.

Crew coordination

The steps that have to be taken to coordinate crews are almost identical to the ones involved with mortared block. The wall crew has to provide correct openings for the carpenters. If any electrical is to run through the block, the wall crew must make cuts and provide conduit for electrical boxes. One builder said that he had his electrician on-site during wall construction to assemble conduit and boxes and show where they should go. Because the wall went up so quickly, he felt that the cost of having the electrician there was minimal. The wall crew has to thread block around any plumbing that comes out of the foundation. It also must put conduit or corner block wherever large lines are to be run. The usual sorts of pre-planning steps taken with conventional block apply here as well.

Connecting to wood

The details of connections were also usually the same as with mortared block. Roof trusses were usually attached with hurricane straps embedded into the bond beam grout. None of the houses we encountered had two floors, so we have no examples of an intermediate floor deck getting attached. According to the manufacturer, it can be done with the same methods used on mortared block.

Windows and doors attached to 2-x bucks that were held to the concrete with bolts or bent nails. You can put the fasteners through the lumber so they stick into the center cavities before the pour. The grout in the center cavities locks them into place. Inside, cabinets and fixtures were attached to the wooden furring strips.

Utilities installation

Small utility lines ran between the furring strips. Some small lines could also go in the inner row of cavities just as they do with most forms of mortared block. An advantage of the Intralock block is that the inner row of cavities is almost completely open because the grout goes in the center cavities. The exceptions to this are the bond beam and the corners (plus any extra reinforcing columns formed of corner block), where utilities usually don't run anyway.

For large lines such as vent stacks, the wall crew can stack up conduit or corner blocks. These can house 4-inch pipe. Reinforcing must either stop at these utility columns or fill in around the pipe.

Change flexibility

Intralock walls are easily moved or changed before the grout pour. All that's involved is unstacking and restacking, along with moving a buck or temporary lintel occasionally. After the grout, you have a very solid wall that has to be cut with diamond blade saws. One builder had to enlarge a door opening. It took about a day to cut the opening on the top and one side, and another day to form a new lintel above and lock it into the rest of the structure.

Manufacturer support

As noted above, Intralock Corporation provides free training to builders and crews who use their block. Our interviewees said they didn't need much more. However, they also said that if they did, the company would send someone to visit them. The company's representatives were described as "very well versed in the system."

Other considerations

The Intralock wall has not yet been tested for sound resistance. However, the great amount of concrete in the wall suggests that it would be higher than the resistance of conventional block.

The builders felt that the job sites where they built with the system were especially neat. There were no mortar drippings or any of the associated materials (sand, mixers) on-site. Instead, a ready-mix truck and pump put just enough grout into the block, then left.

Further information

Intralock Corporation
1001 N.W. 12th Terrace
Pompano Beach, FL 33069
(305)-942-0000

Sparfil Wall System II

Sparfil International, Inc. has developed a mortarless block system that's unusually energy-efficient. It's also made of a lightweight material that works a lot like wood, so striking architectural features are relatively easy. The Sparfil Wall System II is currently available from Gilbert Block in New Hampshire, as well as two plants in Canada. The manufacturer has plans to introduce it in Florida as well. Because nearly all of the houses built of Sparfil in the United States have been in the Northeast, that's where we got our information.

Most of the buyers have been typical of the energy and quality market segments. Interestingly, a few have also been cost buyers who need a basic structure with a higher than normal R-value. The system can provide a high-R, simple wall inexpensively.

The block for Sparfil Wall System II is made of the same concrete-and-EPS bead mixture used for Sun Block (see chapter 5). About 40 percent of its volume is concrete, and 60 percent is foam beads. This gives the material a relatively high R-value. It also makes it possible to cut with ordinary wood saws. One builder told us he found that he could make all his cuts on a small house with an old carbide blade. It would get dull by the end of the job, when he threw it away.

The major materials of the Sparfil Wall System II are pictured in Fig. 6-16. They're block, "Sparbond," and a fiberglass mesh. Sparbond is an elastomeric wall coating that's used with the mesh to form a strong surface bonding. The blocks come in 8-inch, 10-inch, and 12-inch widths. Figure 6-17 shows their exact dimensions. Every builder we spoke with used the 12-inch block because of its higher energy efficiency and because it doesn't cost much more than the others. One, however, switched to 8-inch block for his second floor. We found the 10-inch block used for occasional connection details in 12-inch walls.

There are three types of 12-inch block: a full (16 inches long), a half (8 inches), and a corner (L-shaped). The full 12-inch block is pictured with the other major materials of the system in Fig. 6-16. In the 8-inch size, no corner block is necessary.

6-16 *The major components of the Sparfil system: architectural finish, surface bonding cement, fiberglass mesh, and a 12-inch block.*

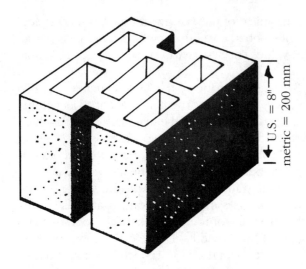

U.S. = 8"
metric = 200 mm

U.S. Blocks
8" (R8 or R15.0) (71.9% solid)

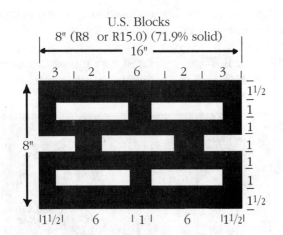

6-17 *Diagrams of Sparfil blocks.*

12" (R12 or R24.5) (61.3% solid)

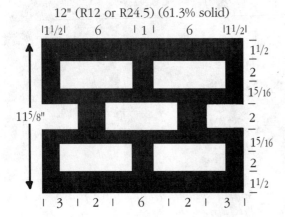

Because the block is easy to cut, the crews can make things like U-block (for bond beams) and conduits (for heavy reinforcing or large utility lines) on-site with ordinary wood tools.

In any of the three sizes, the block has three rows of offset cavities, as diagrammed in Fig. 6-17. The plant will deliver block with optional EPS foam inserts already in the cavities for those who want more insulation than the empty block provides. You can also use the cavities to hold reinforcing bars or utilities, knocking out inserts as necessary to make room for them.

Figure 6-18 shows a diagram of a typical wall section. The first course of block is leveled on the foundation in a conventional mortar bed. The crew stacks all courses above the first without mortar. The manufacturer sizes the blocks to a true 8-x-16 dimension with a grinder at the plant so they stack easily. The builders we spoke with checked for plumb and level frequently when stacking, but said they needed to add shims to correct only occasionally. One carried galvanized roofing nails in his apron and slipped one or two into the joint between blocks where he needed to true one up.

Lintels are formed in one of two ways. The first is to use a standard steel lintel. This has a rib about 4 inches high along the top. To accommodate the lintel without throwing the block above it out of line, you shave the block about a quarter of an inch and cut a groove for the lintel's rib to fit into. The second method is to re-

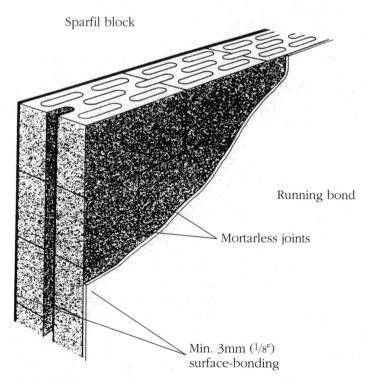

Sparfil block

Running bond

Mortarless joints

Min. 3mm (¹/₈″) surface-bonding

6-18 *Diagram of a typical Sparfil wall.*

inforce a row of block yourself. The manufacturer points out that this has two advantages over a steel lintel: it doesn't break the insulation and it won't corrode. To do it you cut a channel along the center row of cavities to form a sort of thin-channel U-block. You then fill a line of these blocks with rebar and grout. The blocks can be set on a buck or temporary brace above the opening and grouted in place, or grouted on the ground and lifted once the grout has set.

Not much reinforcing is necessary on houses because of the strength of the surface bonding that covers the walls. Basements and other below-grade walls usually require one vertical rebar every 4 feet. Above grade there's usually a bond beam at each floor, but no vertical reinforcing. You can look up the exact amounts of reinforcing to use in the Sparfil manuals, or call the company if it's still not clear for your design.

You can slide vertical reinforcing bars down the cavities after the wall is up, just as you'd do with conventional block. Horizontal reinforcing is handled like the lintel described previously. You cut a channel along the center cavity of the blocks on one course, then fill the channel with rebar and grout.

Once the wall is stacked, it's ready for the surface bonding, which consists of the Sparbond and fiberglass mesh. You trowel on a layer of Sparbond, press sheets of the mesh onto it, then add another layer of Sparbond. The mesh adds strength to the bonding and ensures consistency. It's hard for the crews to skimp on the Sparbond if they have to cover the mesh. It also leaves the surface relatively smooth because there are no loose fibers sticking out of the finished wall.

According to our builders, the bonding on one side (inside or outside) is best done all at once. If you wait too long between layers, the first coat can dry out too much for good adhesion. If you quit halfway down the wall and start again the next day, you can get a seam. So you do each side straight through all around the house, or stop at a break point like a corner.

Labor and materials' costs

The Northeastern builders we talked with compared Sparfil costs with 2-x-6 exterior walls. They said that at current lumber prices Sparfil would be from $0 to $2000 more for a 2000-square-foot (living area) house, depending on the finishes used. When the Sparbond was an acceptable finish, Sparfil's cost was about the same as 2-x-6. If the interior was furred and sheetrocked and the exterior required a screwed or nailed siding, the Sparfil was more. These figures are for Sparfil blocks with the optional inserts in all possible cavities. Because most of the buyers selected the Sparfil system partly for its energy efficiency, they almost always opted for the maximum number of inserts.

Unit costs we were quoted for the 12-inch block ran about as listed in Table 6-3. For a wall of 8-inch block, the total cost is about 60 or 70 cents less per square foot. The block itself is about 50 cents less per square foot, and the inserts to fill it are a nickel less. Plus, the labor is a tad less because the crew can move the lighter units around faster. All other costs are virtually the same.

Table 6-3. Representative Costs for 12-inch Exterior Walls of Sparfil[1]

Item		Cost per Square Foot
Blocks		$3.10
Insulation inserts, all cavities		.70
Sparbond cement and fiberglass mesh		.85
Grout and rebar		.15
Stacking and reinforcing labor		1.30
Surface cementing labor		1.20
Subtotal, structure and insulation		$7.30
Exterior siding and finish, materials and labor		$ 0–3.20
None	$ 0	
Paint only	.50	
Sparfil Architectural Finish	1.00	
Clapboard, with furring strips and paint	3.20	
Interior finish, materials and labor		$ 0–1.50
None	$ 0	
Paint only	.70	
Sheetrock, with furring and paint	1.50	
Total, including finishes		$7.30–12.00

[1] The costs listed here are in the middle of the range of quotes we received from builders reporting their actual project costs.

If the standard Sparbond is an acceptable finish, Sparfil forms a moderately priced wall—about $7.30 per square foot for the 12-inch version, including inserts. It's difficult for a frame wall to be as energy-efficient at this cost. We encountered walls that use this minimum amount of finish (Sparbond only on both sides) in some typical projects directed at the cost market segment (vacation homes in the woods and low-priced rental units). Some weren't even painted (although the manufacturer recommends painting the exterior). The builder just chose the available white-colored Sparbond and considered the result adequate. Inside, the Sparbond is coarser than plaster, but it was smooth enough for these projects. For projects that don't need high energy efficiency or such a thick structure, you can go down to 8-inch block and/or omit the inserts. This should save up to $1.40 per square foot.

Adding extra finishes increases the cost. Nailed sidings (like clapboard) require furring strips. Therefore they cost a little more on Sparfil than on frame. We didn't run into anyone who had used vinyl or aluminum. The Sparfil Architectural Finish is a polymer-modified (PM) stucco available from the manufacturer.

Learning costs

The builders agreed that the rules of the system are easy to learn, but it takes some time for masons and plasterers to adapt. In most cases the builder learned the system well from the manuals and from the manufacturers' representatives, who will come to visit first-timers. Then the builder trained the other crews, or got the manufacturer to do that, too.

Plasterers could pick up what they needed to know after one house. The opinion on masons varied. Some builders thought it took them a few houses to get comfortable with Wall System II.

No one had exact dollar estimates for the extra costs they incurred in their first couple of houses. The time the builder spent was usually a day in training and several hours reading the manuals. Masons were sometimes hesitant to bid because they didn't know how long the block would take them. So before they got used to the system, they might bid their standard labor rates for conventional block, which could add about a dollar per square foot. But most builders assembled their own crew and paid hourly, so they avoided these increased charges. Plasterers didn't generally bid high on the first job. In addition, on the first job the builder had to spend time keeping after the trades to install the product correctly. Most of our builders were somewhere between half-time and full-time on the job site the first house or two, instead of just checking in occasionally like they might on other projects.

The most common problems on the early jobs had to do with leveling. Apparently it's well worthwhile to get the first course of block (the one laid in conventional mortar) as straight and level as possible. One very meticulous builder shot for level all around the foundation within a sixteenth of an inch. If this course is level, the stacking above goes fast because you don't have much correcting to do. But often on a first job, the crew didn't know about this requirement or didn't take it to heart. The first course was a bit uneven and a lot of time got wasted straightening things out up above. Similar problems arose from not leveling the higher courses carefully enough. Although most blocks don't need shims, a few do. When crews didn't check for this and shim wherever needed, they had to do more work on the courses above to compensate.

Buyer reaction

The buyers who have reacted best are those interested in high energy efficiency or general high quality. Buyers focused on energy efficiency have been willing to accept the unusual finishes of a standard Sparfil wall, or else pay the extra for added-on sidings like clapboard. Some buyers of vacation homes enjoyed getting extra-high insulation for about the cost of 2-x-6, and finish was not a high priority for them. A few seemed to like what they considered the "rustic" look of the Sparbond finish. And some buyers selected Sparfil Wall System II for its overall quality.

The reaction of northeasterners in general appears to be an interested curiosity. Blocks and troweled-on finishes are less common there than in many other parts of the country. Some builders told us they got a lot of questions and some interest from passers-by.

Exterior and interior finishes

As previously discussed, the minimum exterior finish is the Sparbond. It looks like stucco. It can be smooth or laid on thick and textured. Buyers who liked this stucco look or wanted to economize on the siding chose to go with the Sparbond only. It comes in gray or white.

The Sparbond can also be painted directly to add color and smoothness at a modest extra cost. Because the surface bonding rarely cracks, a common latex paint should suffice outside. Some builders chose to go to the extra expense of an elastomeric paint.

The Sparfil Architectural Finishes are polymer-modified (PM) stuccoes that produce a variety of troweled finishes in a range of colors. They're available from the manufacturer. Their additional cost is also moderate, and they produce a fancier appearance.

Some of the New England builders we interviewed used conventional clapboard siding, as shown in Fig. 6-19. This required attaching furring strips first, at an additional cost of around $.20 per square foot.

Inside, the builders doing basic homes left the Sparbond uncovered or painted it. The homes built for high-end customers had furring and sheetrock. It's also possible to plaster directly over the Sparbond surface.

Design flexibility

One builder said, "There are very few limitations to this system if you use it to its full potential." The builders we spoke with had used the system for basements plus two stories without problems. The manufacturer says it can go higher, but you would need to check your design with Sparfil International's engineers to make sure it's properly reinforced.

Everyone agreed that unusual angles and openings were easy to make with the system. They take about the same time as doing similar things in frame, and in some cases a little less. To make, say, a 45-degree angle in the wall, you cut a dozen or so block and stack them. As one builder put it, "It's a little time-consuming, but not difficult." One of the builders made curved walls just the way it's done with conventional block. He cut half-blocks at slight angles and stacked them in a stack bond, each one angled a little farther around.

6-19 *Bonded Sparfil wall halfway through siding with clapboard.*

For irregular wall openings, the usual practice was to stack and bond the wall and then cut it in place. One builder stacked up and bonded block to form rough gable ends, snapped chalk lines to mark the roof line, then cut off the excess block. Another formed arches by stacking and bonding his wall, then cutting out the curve with a reciprocating saw. You can use temporary 2-x braces to hold the blocks above the opening in place, instead of wasting a lot of extra block for the purpose. A lintel should go above the top point of the opening. The block below will hang from it with the surface bonding. However, one builder confided in us that he got away without any lintel over small openings.

R-value and energy efficiency

The manufacturer has had a completed 10-inch wall tested with a guarded hot box. It tested at R-18.5 with inserts and R-10 without. Using these numbers as a starting point, the company then calculated what the R-value for the 12-inch block should be. The results were R-24.5 with inserts, and R-12 without. They also calculated values in the same way for the 8-inch block; they were 13.5 with inserts and 8 without.

Finished houses have virtually no complete breaks in the insulation. Some inserts can be omitted and channels can be cut into the block for reinforcing and utility lines. But in these cases, at least part of the block is still intact around the channel, and the block insulates at about R-1 per inch.

The weight of the 12-inch block is about 35 pounds per square foot. For the 10-inch block, it's about 33 pounds, and for the 8-inch block it's about 28. The manufacturer estimates the thermal mass of the 12-inch, 10-inch, and 8-inch blocks at about 7.7, 7.3, and 6.2 Btu/sf/degree Fahrenheit. These figures are higher than the levels required by most energy codes for preferential R-value treatment. However, the system far exceeds most R-value requirements, anyway, unless you're using the blocks without inserts.

Residents agreed that their fuel bills were low in houses built of Wall System II. A vacation house in Massachusetts with a lot of south-facing glass required no heat on sunny days even in the winter. The owners used no heat at all until October. A small house (1500 square feet of living space) in Connecticut with a wood stove had an annual gas bill for heat and hot water of $575 plus two cords of wood.

Thermal mass

As with some other systems, the thermal mass of a Sparfil Wall System II wall is intermingled with the insulation. So it's hard to say whether the mass is inside the insulation envelope or outside it. Nonetheless, occupants noticed definite thermal mass effects. In the Massachusetts vacation home, the heat that accumulated during the day was "enough to get through the night if you didn't mind it being chilly when you got up." From several owners we heard that there were no "cold spots" in their walls, and the warmth was much more even than in other houses. One couple said theirs was the most comfortable house they'd ever lived in.

Water resistance

The builders reported almost no water penetrating to the interior of a Sparfil Wall System II home. The only exception is some leakage through a wall below grade that, according to the builder, had been damaged but not repaired before backfilling. Their opinion was that once the Sparbond is installed, the wall is highly water-resistant.

As previously discussed, block walls that used similar fiberglass-reinforced surface bonding cements have been subjected to the driven-water test and have shown extremely small amounts of water penetration.

Maintenance required

With a finish of unpainted Sparbond, the wall is virtually maintenance free. Some of the builders reported occasionally getting a few hairline cracks in the exterior surface. They were barely visible. The builder usually didn't bother to repair them, and we got no reports of problems arising from them.

If the walls are painted, we suppose that they have to be repainted about as often as other cement products—about two-thirds as often as wood siding. Any added sidings (clapboard, etc.) would require their usual level of care.

Disaster resistance

The fire ratings on Sparfil Wall System II are particularly high. An 8-inch wall survived the fire test for a remarkable four hours without a failure of any kind. In a separate test, an 8-inch wall with an added load on it survived two hours without a failure. The company hasn't done additional tests to find out how much longer the wall could go before it does fail, or to find out how much better the 10- and 12-inch block would do.

To date, houses of Wall System II have generally not been built in high wind or earthquake areas. However, according to the manufacturer, they could be built to withstand these conditions by adding reinforcing. Sun Block, which is made of the same material, is regularly used in houses in Florida that had to meet design loads of 110 mph. One builder speculated that the surface bonding of Wall System II would hold the house together even better than mortared block walls.

Product availability

Currently the materials are available in the United States from Gilbert Block, which ships them from Hooksett, New Hampshire. They're also available from two plants in eastern Canada. The manufacturer has plans to offer it in Florida as well some time in the future.

To date, U.S. builders have used the product throughout New England and on Long Island and in upstate New York. Most ordered all of the special materials (block, Sparbond, fiberglass mesh, architectural finish) from Gilbert Block, even though some of them can be bought elsewhere. They ordered a few days to two weeks in advance and say they always got the shipment on schedule. Shipping materials for a typical house to the most distant locations (New York State) adds a few hundred to a thousand dollars to the cost.

Labor availability

Most of the builders hired unskilled workers and ran the wall crew themselves. They usually had difficulty getting ordinary masonry crews to adapt to the system. The work involves shimmed stacking instead of mortaring, and sometimes a lot of measuring and cutting. These are things masons don't normally do. We were also told that the work of masons was often a little slow and of uneven quality, especially to start. On the other hand, everyone agreed that the heavy tasks are easy for an unskilled worker to learn and to do correctly. So the builder himself would learn the technical details and supervise unskilled workers, who did most of the legwork.

A couple of builders had great luck with masons from Europe or Latin America. They were often more familiar with lightweight concrete systems, and they understood the reasons for the different procedures used with Sparfil Wall System II. They could be trusted to run a crew with much less supervision from the builder.

Plasterers generally adapted fine. One builder said it was worth giving them a few hours of instruction in advance, but all agreed it took them only one project to get really efficient with the system. Some builders did their own surface bonding. They said it was easy to learn, and they were satisfied with the quality they could get even with little training. One said he got satisfactory results, despite not being an expert on the trowel, by smoothing the Sparbond with a 6-inch paintbrush while it was still wet. Gilbert Block offers to send a crew to do the walls for anyone who can't find one. We didn't hear of any builders who resorted to this, however.

Building department

None of our builders had any difficulty getting building department sign-off on the system. According to one builder, the company has "a flood of good technical documentation," including evaluation reports from BOCA (number 85-11), SBCCI (85-53), the Department of Housing and Urban Development (approving the system for use in HUD-sponsored projects, Materials Release No. 1249), New York City (MEA280-93-F), and other local and regional codes in the United States and Canada. When the building departments asked, the builders submitted these materials, and that was enough. In some cases we heard about, the inspector didn't even bother to look at the structure until after the Sparbond was on. This makes some sense because the Sparbond binds the structure together. That can save one inspection visit (the one for framing and fire blocking).

Required calendar time

The builders claimed that construction is faster than with frame if you use the Sparbond alone for the finishes. If you add sheetrock and an exterior siding, it's about the same.

One put up the exterior walls for a two-story, 3300-square-foot (living space) home, including all surface bonding and openings, in one week with a five-person crew. Remember that if you don't want siding or sheetrock, this work provides the structural wall, all the wall insulation, and all the finishes.

The speed of Wall System II comes mostly from the stacking. This proceeds at about 500 blocks per worker per day, versus about 250 per mason with mortared block. Surface bonding ran two to three days of actual work, depending on the size of the crew and the size of the house.

Some jobs were delayed by crew changes. These were always when a builder was doing his first Wall System II building. When the original crew didn't work out, the builder lost a few days bringing in a new one and, sometimes, fixing the work already done.

Stacking can proceed in any weather. The manufacturer recommends that the wall be covered on top to protect against rain and, if it does get soaked, that it be allowed to dry for 48 hours before applying Sparbond. Surface bonding can't be done in the rain, and it should be protected from rain for a few hours after it's up. You can apply Sparbond in dry weather, but you should wet it down for a day or two afterward so it cures properly. It shouldn't freeze for 48 hours either, so you need to time the application accordingly.

You can start working on the other parts of the house immediately after the wall is up. Some builders even added floors or roofing before they did their surface bonding. But anything that could apply a side load to the walls, such as backfilling, has to wait until a couple of days after the Sparbond is on.

Crew coordination

The only special coordination necessary was for the wall crew to provide for things that go through the wall later. This can include reinforcing, utilities, and sometimes floor joists. For vertical rebar, the crew has to punch the inserts out of the column of cavities to be reinforced. For horizontal rebar, the workers cut a channel along the top course of block. They might also have to wait for the grouting. But because the channels are small and therefore require little grout, the wall crew often mixed up its own grout on-site and did the reinforcing.

For electrical lines, the crew usually routs out channels in the surface of the wall before surface bonding. The workers might also insert conduit, depending on local code and preferences. Some builders had the electrician out to show where the lines should go, but most simply had the electrician explain it in advance and relayed the information to the wall crew. Floor decks require grouting J-bolts into the block or cutting out pockets.

Because the builder himself was the field supervisor in almost every case, these things presented no problems. The builder knew they were coming and why, so it was easy for him to direct the workers to include them.

Making connections

Connections of floors and roofs are covered in detail in the system manuals. For floors, there are three alternatives given. One is to grout joist hangers into pockets cut into the block. The others depend on having a shelf formed by going from a thicker to a thinner block. Some builders switched from 12-inch block to the 8-inch or 10-inch block for one course to form a shelf for the floor deck. With a 4-inch shelf you can rest a 2-x-4 plate flat. With a 2-inch shelf you rest a 2-x header on the shelf and fasten it to the block behind with J-bolts grouted in. The joists attach to the header with joist hangers. Roofing lumber attaches to a top plate. This is fastened to the top of the wall with J-bolts grouted into the center cavities of the block below.

The builders claim that these details are generally pretty easy. The amount of grout involved is small because the cavities are narrow. And you usually don't need to put paper or felt below the cavities to be grouted because the inserts below will stop the grout flow.

Other connections were almost as easy as with lumber because Sparfil takes fasteners readily. Some builders used cut nails to connect interior walls. Another used ordinary wood nails, driven at different angles. He preferred not to nail straight in because the material doesn't "grip" fasteners quite like wood. With vibration, a nail can pull out. For cabinets and fixtures, most chose to use plastic screw anchors and wood screws. Holes for the anchors went fast with an ordinary wood drill. For furring strips, the preferred method was to glue the strips and hold them while drying with sheetrock screws shot directly into the Sparfil.

Windows and doors can attach in most of the same ways they go onto conventional masonry. Most of the builders used a buck and screwed the window or door to that. Some attached the buck to the block with plastic or lead anchors and wood screws. One used the cavities in the block to hold the buck firmly. If you look at Fig. 6-16 closely, you see that the block has a "half cavity" running like a channel up each end. Its purpose is to form a full cavity when the block is stacked beside its neighbor, to hold another insert. But when you stack the block around a window or door opening, these half cavities line up on either side. Figure 6-20 shows this. With 12-inch block, you can get a 2-x-4 into each of the resulting "channels" in the jambs, with two more across the top and bottom to complete the buck.

6-20 *Openings in a Sparfil wall after surface bonding. The channels along the jambs are available to hold the sides of a buck made of 2-x lumber.*

Utilities installation

If the walls are furred inside, small electrical and plumbing lines go between the furring strips. If there's no furring, the wall crew generally routs out a channel for small lines in the surface of the block after it's up. Ordinary wood routers can do the cutting. Figure 6-21 shows a detail of this for wiring. After routing, the crew inserts conduit in the channel. The crew also cuts a hole for each box and holds it temporarily in place with ordinary wood nails driven into the block. The channel is filled in with grout, and then it and the seams around the boxes get covered over by the Sparbond. This locks everything into place.

Plumbing can be handled the same way, and big lines can go in the cavities. For very large lines, you can cut out the concrete between two cavities, forming one big one. But the builders avoided running plumbing in their exterior walls, and the manufacturer recommends against it.

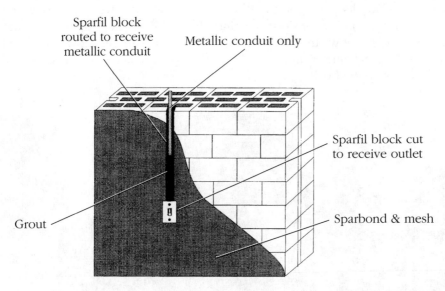

6-21 *Detail of electrical box and cable placed in openings cut in a Sparfil wall.*

Change flexibility

Changes are generally easy with Wall System II. The builders sometimes came across errors (unlevel blocks, misplaced openings) while stacking. These things were rapidly corrected with a few shims or some restacking of block. After Sparbond was applied, most changes were still easy because the block cuts readily. Builders cut new openings and moved old ones with reciprocating saws. If the new openings were small, some didn't even add a lintel. When it was necessary to add or move a lintel, more work was involved because the crew had to cut out the block above the opening and replace it. But it was usually possible to get away without much bracing because the surface bonding held the blocks above. One builder who had to move a window opening, lintel and all, guessed that it might have cost as much as a couple hundred dollars, mostly for labor.

Manufacturer support

Everyone agreed that the written documentation on Sparfil Wall System II is excellent. Some builders claim they practically taught themselves how to use the system with the manuals.

Gilbert Block sends at least one representative to the job sites of builders who haven't used the block before. The reps might stay there up to a day answering questions and demonstrating anything that's unclear. The builders we talked with appreciated this, although sometimes they already understood much of it from the manuals. The visit was often more help to their crews, who hadn't done as much homework. Gilbert reps will also visit a building department before it receives plans on a new house. They educate the officials about the system so they aren't surprised by it. All of this personal help will probably be harder to get for a project that's far from Gilbert's New England base, however. People at Gilbert Block and Sparfil International headquarters in Canada are also available to answer questions by phone. Sparfil International has an engineer on staff to help with structural or design questions.

Gilbert claims now that it will guarantee a price on the walls of a Sparfil house. The builder can send his plans, and Gilbert will give a quote for all materials and labor that go into the block and surface bonding. If the builder can't find a crew that will build for that price, Gilbert will send its own to do the work. We did not encounter a builder who had taken the company up on its offer, however.

Other considerations

We're told there's no significant mess to the site from Sparfil Wall System II. According to one builder, "You get these loose foam beads when you cut. They blow away until you have the walls up. Then you just sweep them off the floor." There's no mortar and little grout. The builders said the Sparbond generally doesn't drip or splatter.

The blocks have the advantage of very low breakage. Even when hit hard, they don't generally crack or shatter. They just dent. We were told that waste from breakage is near zero, versus 3–5 percent for some conventional block projects.

Further information

Sparfil International Inc.
376 Watline Avenue
Mississauga, Ontario
Canada L4Z 1X2
(416)-372-6853

Del R. Gilbert & Son Block Co., Inc.
RFD 2, Province Road
Laconia, NH 03246
(603)-524-1353

7

Poured-in-place systems

In conventional poured-in-place construction, a crew erects forms of plywood or steel that make a "mold" in the shape of the desired walls. They put rebar inside the forms, suspending it in place on metal or plastic rods or wires. Then they pour concrete inside. After the concrete hardens, the crew strips the forms to leave the reinforced concrete wall.

What's new in poured-in-place construction is the stay-in-place systems. With these systems, you build the form wall out of foam panels or blocks. After pouring and hardening, the form work is not stripped away; it's left in place. This foam, which once acted as the form, later serves as insulation, exterior wall sheathing, and, in some cases, wall studding.

There are more than a dozen different stay-in-place systems available in the United States, each with a different set of features. Table 7-1 summarizes the key characteristics of the ones we're aware of. We group them into two categories: panel systems and grid systems. The stay-in-place panel systems use off-the-shelf sheets of foam insulation. Pairs of foam sheets are held apart with plastic ties that keep the spacing even and hold the foam against the force of the concrete during pouring. The stay-in-place grid systems use specially molded foam "blocks" that are mostly hollow in the center. The crew stacks them up and pours concrete down the hollow core.

Before describing the two types of systems in depth, it's useful to review some of the terminology of conventional poured-in-place concrete walls. The concrete itself comes in various compressive strengths. This is sometimes loosely referred to as the *load-bearing strength*. It's how much load or weight the concrete can support without cracking. In residential construction, most of the concrete has a compressive strength of 2500 or 3000 psi (pounds per square inch). Other strengths might be required for special purposes, and the higher strengths cost a little more. The concrete also comes in different sizes of *aggregate* (the stones included in the mix). Most concrete in residential construction is "¾ inch". This means that the largest stones measure ¾ of an inch across. Con-

Table 7-1. Characteristics of Stay-in-Place Forming Systems

Brand Name	Manufacturer	Unit Dimensions (width×height×length)	Screwing Surface	Interlock Edges	Form Cost Per Sq. Ft.	Filled by One Yard of Concrete	Approx. Stated R-value	Evaluation Reports	Notes
Panel Systems									
Lite-Form	Lite-Form Inc.	12" × 8" × 8'	Yes	No: ties connect panels	$2.10–$2.25	40.5	20		
Polycrete	Distribution of Polycrete of Montreal, Inc.	13" × 12" × 8'	Yes	No	$2.50	40.5	21.5		
R-FORMS	R-FORMS, Inc.	8" × 4' × 8' to 16" × 4' × 8'	Yes	No: brackets connect panels	$1.80 plus $.61 for bracing (if needed)	40.5 (for 8" concrete wall)	20	BOCA 91-54 ICBO 3628 SBCCI 8965 SBCCI 8965	Available in several widths
Grid Systems									
AAB	AAB Building Systems	11.5" × 16.75" × 48"	Yes	Yes	$2.73	48	26	In process	Canadian regional distribution with specialty products available
Energrid	Energrid	10" × 15" × 8'9" to 12" × 30" × 10'	No	No	$2.28	100	30		Available in various sizes; forms made of mixture of concrete and EPS beads
Featherlite	Featherlite Inc.	8" × 8" × 16"	No	Yes	$3.00	Variable, about 200	30.4	ICBO 4643	Partly filled; concrete used depends on number of cavities filled
FFCF (Foam Furred Concrete Form) Panel	EnerG Corp., Inc.	8" × 2' × 8'	Yes	Yes	$2.53	Variable, about 80	27	ICBO in process	Partly filled; concrete used depends on number of cavities filled

GREENBLOCK	GREENBLOCK Worldwide Corp.	10" × 10" × 40"	Yes	Yes	$2.88	30.5	20		
ICE Block	ICE Block Building Systems	9.25" × 16" × 48" or 11" × 16" × 48"	Yes	Yes	$2.16	62 (9.25" block) or 53 (11" block)	28-32	ICBO 50-55 BOCA 92-28 SBCCI in process	
Keeva	Keeva International, Inc.	8" × 12" × 48"	No	No	$1.50	Variable, up to 208	20	ICBO in process	Partly filled; concrete used depends on number of cavities filled
Reddi-Form	Reddi-Form, Inc.	9.625" × 12" × 48"		No	Yes	$1.56	80 19.5	ICBO in process	BOCA 90-81
Smart Block	American ConForm Industries	10" × 10" × 40" and wider	No	Yes	$2.70-3.80	55 (10" block)	22	ICBO 45-72 BOCA 89-67 SBCCI 94-26	Available as 10" block with EPS webs or in variable widths with slide-in plastic ties
3-10 Polysteel	American Polysteel	9.25" × 16" × 48" or 11" × 16" × 48"	Yes	Yes	$3.00	72 (9.25" block) or 53 (11" block)	20	BOCA 90-12 ICBO 4295 ICBO 3401 SBCCI 9342	

crete with smaller aggregates is a little more expensive. It's used when the mix has to flow well through small spaces, such as when it's pumped into the forms with a concrete pump, or when there are obstructions inside the forms that the concrete needs to get through.

Concretes also differ in their *slump*. This is how thin or "runny" the concrete is. The ready-mix suppliers make the slump higher mostly by adding water. You measure the slump of a batch of concrete by packing some into a 12-inch-high pyramid-shaped mold, turning the mold over (sort of like molding sand in a bucket for making the tower of a sand castle), and seeing how far the concrete slumps down. If it drops down 1 inch, the concrete is said to have a 1-inch slump, and so on. Typical slumps for concrete used in residential form work are 4–5 inches, although the crew might call for a higher slump if they need the concrete to run into small cavities. However, too much water increases pressure on the forms (which increases the risk of a blowout), and makes the final concrete weaker. Occasionally contractors have special agents such as fiberglass or pigments added to their concrete to give it particular properties such as extra strength, color, or water resistance.

Amounts of concrete are measured in yards, which is really slang for cubic yards. Figuring the amount needed for a specific job is second nature to forms crews. But it can be tricky for trades that normally work with feet and don't often calculate volumes. (There are 27 cubic feet in one cubic yard.)

Conventional forms, like the stay-in place panels, consist of sheets (usually of plywood or steel) that are held a constant distance apart by ties. The ties are pieces of wire that hold the panels against the weight of the concrete during the pour, but that the crews can break away afterward. The forms are also coated on the inside with form oil or some other release agent that makes it easier to remove them after the concrete hardens. The forms are also braced with various brackets and 2-× lumber, which can also be adjusted as necessary to get the form work plumb.

When the forms are nearly set and the crew is double-checking them, someone calculates how much concrete is necessary to fill them, adds 5 percent or so to be safe, and calls in an order to a concrete company. The pouring of the concrete (also called the pour) can be arranged so that the crew fills one section of the wall to the top, then moves down to the next section and fills it, and so on all the way around. But this puts a lot of pressure on the forms, especially at the bottom of the wall. Instead, it's common to fill the forms only part way up all the way around, then go around again and fill up a little farther. This allows the concrete at the bottom to harden some before the form work there has to bear the full weight of a totally filled wall. One of these times around is called a *lift*. A typical lift fills the forms up 4 feet high, just enough to complete an 8-foot wall in two passes.

In most pours, the driver positions the ready-mix truck near the wall, suspends a chute that's attached to the back of the truck over the open wall cavity, and releases concrete to slide down the chute and into the wall. The forms crew directs him on when to put how much concrete where. By moving the truck a few times and swinging the chute from side to side, the concrete is made to drop all around the perimeter. When the top of the wall is higher than the truck or it's important to pour at a slow, controlled rate, the forms crew usually uses a concrete pump. When a pump is used, the ready-mix truck pours its load into the hopper of the pump. The pump then forces the concrete down a hose that the crew holds over the wall to drop the mix into place. Independent pump operators charge about $500 to bring their equipment out and run it to fill the forms for one floor of a typical house of stay-in-place panels. You also have to order a special form of concrete designed to flow correctly, called a *pump mix*. It usually costs about 2 percent more than a standard mix.

Especially if the concrete is poured fast or the form work has weak points, the forms can burst open somewhere. This is a *blowout*. With experienced crews, blowouts are rare, but the crews prepare for them anyway because the consequences can be serious. Concrete flows onto the surrounding ground and begins to harden. The section of wall that blew out has to be rebuilt, but the forms there are broken. The pour has to stop while the problem gets corrected. The crew usually has extra forms or sheets of plywood and extra bracing that they can put on top of or in place of the blown forms so they can reinforce the damaged wall section and start pouring again. The concrete that spilled out is shoveled back in or goes to waste. A few days after the pour the concrete is hard enough to backfill (if it's below grade) or support a heavy load.

In the following section, we describe the stay-in-place panel systems first. They're more similar to traditional forms, so they're a little easier to understand. While describing them, we explain some basic concepts that apply to all the new stay-in-place systems, which we refer back to when we describe the grid systems. If you're interested in the grid systems, you might want to read about the panel systems as well to get this background material.

Stay-in-place panel systems

There are two established panel systems in the United States: Lite-Form and R-FORMS. A third from Canada, Polycrete, has recently also become available in the United States. We were not able to interview U.S. builders about Polycrete before the publication of this book, however.

The term "panel" refers to both the forms themselves and to the shape of the concrete in the final wall. The forms are made of standard extruded EPS foam board, typically a 2-inch-thick sheet outside the wall and another one inside, for a combined 4 inches of foam. Plastic ties hold the two sheets an even distance apart so that they'll form a flat concrete wall of constant thickness.

If the EPS were stripped off the final wall, you'd see a concrete wall just like you get from traditional forms. But unlike the traditional systems, the foam form is internally secured to the concrete by the ties and left in place. There it provides insulation and backing for finishes.

The ties of both systems come in different sizes to produce a concrete wall of different widths (4 inches, 6 inches, 8 inches, or 10 inches). R-FORMS' ties can also use foam sheets from 1 to 2 inches, but typically are set for 2. Lite-Form ties are designed exclusively for 2-inch sheets. Both companies' ties also have a flat plastic end inside and out that will hold a screw, allowing you to attach both interior and exterior finishes, and some trim and fixtures, directly to the wall without furring strips. The ties fall every 8 inches horizontally and vertically with Lite-Form, and they fall every 16 inches with R-FORMS. So both the panel systems provide forming, insulation, and studding in one step.

The standard package of R-FORMS materials includes 4-feet-x-8-feet flat sheets of off-the-shelf foam and a variety of plastic ties. The crew drills holes in the foam and sets ties in the holes to form panels that sandwich an air space between two foam sheets, as shown in Fig. 7-1. On the footing, they snap chalk lines to mark the exact position of the outside and the inside of the wall. Then they nail "guides" into the footing. These are usually 2-x-4s nailed to the outsides of the chalk lines. Then the crew can set the panels snugly between the guides, and the guides brace the panels along the bottom edge. Horizontal rebar rests on the ties, and the verticals are wired upright, just as they are with conventional forms.

The panels can be set sideways to form a 4-foot wall, stood up to form an 8-foot wall, or stacked to go up to greater heights. Because they don't have to be filled all the way, you can form a wall of any height in between, too. R-FORMS sells special plastic channels to connect adjacent panels and connect

7-1 *Assembling R-FORMS panels on-site.* R-FORMS, Inc.

door and window bucks to the panels. The diagram in Fig. 7-2 shows some of these. There are also special metal brackets that you use to hold 2-× lumber tight against the form wall to brace it. Some such brackets are also in Fig. 7-2. The manufacturer says that there is now a new bracing system that employs metal studs and eliminates the need for the metal brackets. Figure 7-3 shows an erected wall braced with the new metal stud system.

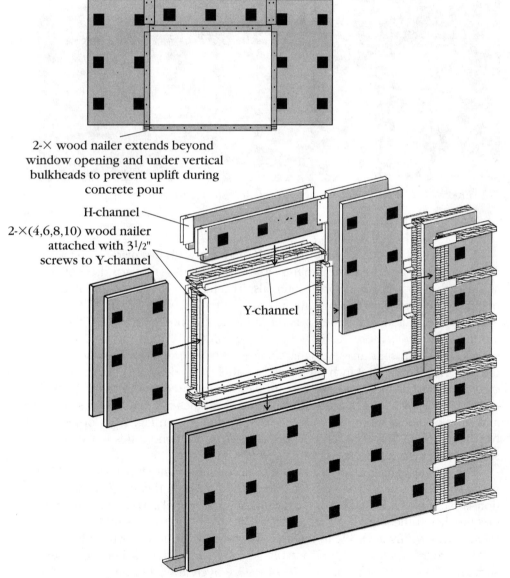

2-× wood nailer extends beyond
window opening and under vertical
bulkheads to prevent uplift during
concrete pour

H-channel

2-×(4,6,8,10) wood nailer
attached with 3¹/₂"
screws to Y-channel

Y-channel

7-2 *Diagram of pieces in a typical R-FORMS wall section.* R-FORMS, Inc.

7-3 *Erected R-FORMS wall braced with the new steel stud bracing system.*
R-FORMS, Inc.

Lite-Form is a little different. Although the form walls are made of standard foam sheet, the sheets are cut into 8-inch-x-8-foot "planks" that are also notched along each 8-foot edge at 8-inch intervals. The ties are made to slide into the notches. The crew puts ties into the notches of two planks of foam to form an 8-inch-x-8-foot sandwich panel, sets these inside the footing guides all the way around, then slides more ties into the notches on top of the panels. Figure 7-4 shows how the planks are set so that their notches will align for the insertion of the ties. This leaves an 8-inch-high wall of panels with ties on top of them. The crew then pushes more pieces of foam over the exposed ties to build the wall up another 8 inches, puts on more ties, and continues alternating rows of ties and rows of planks all the way up the wall, as depicted in Fig. 7-5. Corners are held together with special "corner ties." Lite-Form uses more ties but less outside bracing than R-FORMS.

For a little extra money, the Lite-Form company will ship pre-assembled wall panels that can go up to the total height of your wall. If you use these, erection is more like R-FORMS. You stand up the large panels. You attach them to one another at the joints with wire that you loop around the ties in the adjacent panels. This wiring is visible in the completed Lite-Form form work in Fig. 7-6.

Polycrete uses 12-inch-x-8-foot planks of foam with ties between them, something like Lite-Form. Figure 7-7 shows a Polycrete wall in progress.

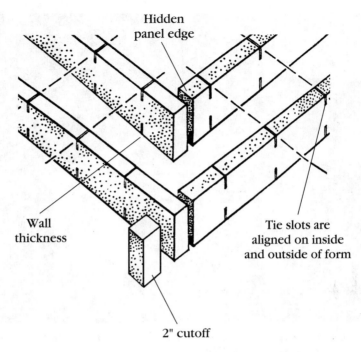

Hidden
panel edge

Wall
thickness

2" cutoff

Tie slots are
aligned on inside
and outside of form

7-4 *Alignment of planks in a Lite-Form wall.* Lite-Form, Incorporated

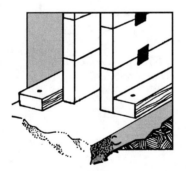

7-5 *Diagram of a Lite-Form
wall section showing the
alternation of planks and
ties.* Lite-Form, Incorporated

With either system, builders tell us you should make the footing as level as possible. The large 4-foot-x-8-foot panels are somewhat forgiving if placed on uneven footings, at least more so than the grid systems are. However, if you build up more than one level, the lower level of panels will have to be trimmed to true so you can attach the second one.

The manufacturers suggest 3000–3500-psi concrete with a 4–6-inch slump. The more experienced builders we talked with said that they could pour an 8-foot wall in one lift, but recommended that it be poured in 3-foot to 4-foot lifts (which is also what the manufacturers recommend). The pour in Figure 7-3 is

7-6 *Completed formwork for a Lite-Form wall, including a curved section and wiring tying adjacent panels together.* Lite-Form, Incorporated

7-7 *Formwork for a Polycrete wall.* Polycrete Industries, Inc.

being done with a conventional chute, and the one in Fig. 7-8 is being done with a pump.

When a blowout occurred, the pouring stopped, just as it does on a conventional job. But with a panel system, the repairs are faster. The concrete in the blown section is cleaned out, and the foam put back in place. The crew then puts a square of plywood up to the blown section, and another one up to the opposite side of the form in the same place, runs threaded rod through the wall and through predrilled holes in the two opposing sheets of plywood, and screws down on the rods until the plywood holds the foam to just the right thickness. If all the materials were prepared in advance, this process took all of ten to fifteen minutes before the pour could start again.

7-8 *Pumping concrete into a Lite-Form wall.* Lite-Form, Incorporated

The panel systems appeal to quality buyers and energy buyers, providing them with a stronger wall and higher R-values for a couple thousand dollars more than frame. Some builders who have learned to use the systems in high volume and driven down their costs sharply have also had success selling to value segment buyers, particularly in the Southeast. The components are readily shipped, and the manufacturers have sent them to all corners of the country.

Labor and materials

Builders told us that an average 2000-square-foot (of living space) house built of a panel system would cost about $2000–3000 more than 2-x-4 frame, if the siding materials were the same. This varied widely by region, however. In some southern states, where the materials costs of concrete and stucco and the cost of labor to build the walls are all relatively low, the panel systems were sometimes close to the same price. This was especially true for high-volume builders who had worked to increase the efficiency of their stay-in-place panel construction. One Florida builder using R-FORMS strips the outside foam off the finished wall and reuses it as the interior for the next house. This saved money on foam and made his siding cheaper because he could put stucco onto concrete (which requires no mesh) instead of foam.

Table 7-2 gives costs per square foot that were in the middle of the range we got from builders around the country. The cost of the form materials, including a typical shipping charge, was about $2.25 per square foot of gross wall. With Lite-Form, this figure includes the foam and ties. With R-FORMS it's partly foam and ties (about $1.50 of the total $2.25), with the rest for brackets. The special brackets that R-FORMS uses to secure 2-x-4 lumber against the foam for bracing can be reused, just like conventional forms, so technically they're an equipment cost, not a materials cost. As with any piece of equipment, it's difficult to say exactly how much of the cost should get counted into any one job. In speaking with builders, $.75 proved a reasonable average. However, some small-volume builders figured the cost of their brackets at each job to be as much as $1.10. Builders actually spent about $2.00 per square foot for the brackets the first time they used R-FORMS, then they bought additional or replacement brackets as needed. The manufacturer tells us that with the new bracing that uses metal studs (instead of brackets and 2 × 4s) the total cost of foam and ties reduces to $1.90.

Table 7-2. Representative Costs for 8-inch (Concrete) Exterior Walls of Stay-in-Place Panel Systems[1]

Item		Cost per Square Foot
Structure and Insulation		
Foam, ties, and (for R-FORMS) brackets		$2.25
Concrete		1.00
Rebar		.25
Lumber		.25
Labor		1.30
Subtotal, structure and insulation		$5.05
Exterior siding and finish, materials and labor		$2.00–4.00
PC stucco and paint	$2.00	
PB stucco	4.00	
Clapboard, with furring strips and paint	3.70	
Interior finish, sheetrock and paint		$1.50
Total, including finishes		$8.55–10.55

[1] The costs listed here are in the middle of the range of quotes we received from builders reporting their actual project costs.

Shipping charges are $1.00–1.50 per mile for the entire load. So if you're next door to the plant or distributor, your total materials bill could be about $100 lower than our figures suggest, and it could be a hundred or so more if you're especially far away.

A typical above-grade wall has 6 inches of concrete (for a total thickness of 10 inches, including foam). At $55.00 per yard, this comes to a cost of about $1.00 per square foot for the concrete. Rebar placed every 4 feet both horizontally and vertically (a typical amount above grade) adds another quarter, as does the lumber for forming the bucks around window and door openings.

The exterior finish was often some form of stucco or a wood siding, and occasionally vinyl, aluminum, or a brick or stone veneer. For wood it's usually necessary to add furring strips first at a cost of about 20 cents per square foot. Because there are screwing surfaces built into the walls, no strips are necessary for screwed sidings such as vinyl and aluminum. So they should go onto the panel systems for about the same cost as they go onto frame.

Learning costs

From the builders we spoke with, we learned that after their third home most of the uniqueness of the system was worked out, and construction proceeded as smoothly as on conventionally framed homes. The builders (most of whom had an experienced forms person on their crew) said that the learning costs were small. There are a few tasks that go slower at first, however, because the crew has to get the hang of them. Because the foam isn't as stiff as traditional form work, there's extra time spent in bracing. There's also a lot of cutting of foam, but this is light work, and it's easy once you have practice.

Blowouts, when they occurred, were also handled a little differently, as described previously. The experienced crews kept a few predrilled, 2-foot-×-2-foot or 3-foot-×-3-foot sections of plywood and some threaded rods, washers, and nuts onsite in case of a blowout. They generally agreed that handling a blowout is easier with foam than conventional form work, but you have to learn how to do it.

Electricians, plumbers, and HVAC subcontractors need to specify before the pour where they'll be putting lines through the walls so sleeves can be properly placed. This isn't difficult, but it involves preplanning that some of them aren't used to. They also have to learn to use either a hot knife or a router to cut channels into the foam for any lines they want to run inside the exterior walls. Again, most decided that it's easy, but it's different. Overall, builders indicated that you might expect to spend an extra $1000–2000 on your first stay-in-place panel home, mostly for labor.

Buyer reaction

The buyers' reaction is in part dependent on the area of the country. In the Northeast and Northwest, these homes were built for specific customers who were interested in the reduction of noise, the insulation, and the solidity of a concrete home. The public in general in those areas regarded the use of concrete as a novelty. In the Southeast and Southwest, where concrete homes have been built for decades, the panel houses were considered just a variation on an old theme. People were generally impressed by the solidity of the system when a builder described it to them, but if no one mentioned it they usually didn't notice that the house was different. In the states where wind loads are a factor, the system attracted buyers for its reduced insurance costs and the peace of mind it provided.

Exterior and interior finishes

Stucco is the exterior finish most often used. It's usually applied by scratching the EPS panel with a wire brush or flooring-grade sandpaper and applying a base or binder coat and then the finish coat. Some PC and PM stucco manufacturers suggest that a mesh be applied first to hold the coats on securely. If a mesh is to be applied, you need some type of fasteners to hold it to the wall. One way to handle this is to push wires through the outside layer of foam before the pour. After the pour, they're locked into place and serve as tie wires for the mesh. An alternative is to install concrete fasteners such as powder-actuated pins or concrete nails after the pour by shooting through the foam and into the concrete. If you use a PB stucco, installing fasteners is usually unnecessary. With them, a fiberglass mesh goes on the surface of the foam as part of the system. The cost of applying any of these stuccoes, listed in Table 7-2, is a little lower than putting them onto frame because the foam backing is already in place. It's higher than stuccoing over bare concrete because you have to add a mesh of one type or another.

Because the panel systems both use a tie whose exposed head acts as a screwing surface, if you're applying a screwed siding (vinyl, aluminum) you can fasten it directly to the ties. Builders using a siding that must be nailed (clapboard, shingles) screwed furring strips to the ties and then nailed the siding to the strips. The furring adds $.20 or so per square foot to your siding cost.

For the interior, we heard about a number of different methods for installing sheetrock. The most common was simply to screw it to the tie ends. Some builders also applied glue for extra strength. Others attached furring strips to the interior and then screwed the sheetrock to them. We were told the benefit of doing this was that it provided an especially true wall, and the electrician and plumber could run their wire and pipe between or through the strips. A few builders routed out a channel of the foam at the top of the wall, set a furring strip in it flush with the foam, and fastened the strip to the concrete to act as a nailer to secure the sheetrock along the top.

The experience of our builders was that the ties sometimes deformed slightly under the internal pressure of the concrete during the pour. The edges of the ties pushed out, deforming the tie head into a shallow cup. In order to remove this cupping before applying finishes, one builder had someone go around the structure and grind down the tie heads with a hand-held grinder or sanding disk.

Below grade, almost everyone used damp-proofing. You have to watch for the compatibility of products. Some petroleum-based damp-proofers can eat away the foam. A latex-based emulsion doesn't damage the foam and is available from large contractor supply houses.

Design flexibility

The builders we spoke with used the panel systems either above or below grade or both. Below grade you simply need to reinforce the concrete as you would any poured concrete foundation. The builders who used it there also stressed the ease of finishing the basement for relatively little additional cost. You simply add sheetrock inside. Insulation and a measure of waterproofing are already installed.

For those builders who used the system above grade and stuccoed the exterior, the many finish options added design interest. The addition of architectural detail (such as raised trim, quoins, and so on) was as easy as nailing or gluing on additional pieces of foam board, then stuccoing over everything.

Irregular corners and bays require about the same amount of labor as they do with wood frame. Workers cut the foam by hand and piece it together with tape and/or wire. They sometimes added additional bracing at the corners, too. Curved walls are made by cutting the forms vertically at an angle every few inches, to get a series of small angles. With Lite-Form, you turn the planks sideways so they stand up vertically for the curved section. You can then simply trim each 8-inch plank. This is what was done with the curved wall in Fig. 7-6.

Irregular openings are relatively easy to make. You cut the foam either before erecting a panel or after it's up, build a buck of the desired shape, and glue or nail through the foam to hold the buck in place or (with R-FORMS) clip it on with plastic channel. The trickiest part is making an accurate buck.

R-value and energy efficiency

The energy efficiency of the panel systems far exceeds most conventional construction. Both Lite-Form and R-FORMS normally use two layers of 2-inch extruded EPS foam with an R-value of approximately 5 per inch. This makes a total insulation R-value for the wall of 20. The manufacturers haven't calculated any more sophisticated form of R-value for their finished walls, but the insulation R-value is likely to be relatively accurate in this case. The insulation is of even thickness across the entire wall and isn't regularly broken by anything except the ties. And because the ties are plastic, they should conduct only small amounts of heat. To get a nominal wall R-value, you would need to add about R-2 for the concrete.

Depending on your particular building methods, there could be several additional minor breaks in the insulation: tie wires for stucco mesh, screws to attach furring strips, sleeves for plumbing and electrical cables, and channels cut into the inside foam for electrical and plumbing lines.

Note that R-FORMS is designed to let you use different thicknesses of foam, and both systems can use foams other than extruded EPS. The most common substitute was 2-inch molded EPS bead board, which is cheaper but not quite as strong. Because different foams and different densities of the same foam have different R-values, which one you use will affect the R of the final wall.

The thermal mass of the walls depends on the thickness of the concrete. With a 6-inch wall, which is common above grade, the concrete normally has a weight of about 70 pounds per square foot and a thermal mass of almost 16 Btu/sf/degree Fahrenheit. These figures are far in excess of the minimum requirements to get special R-value treatment under most energy codes. However, panel system walls almost never need an R-value credit because their uncorrected R-values far exceed most codes.

The manufacturers haven't calculated mass-corrected R-values for their walls. But judging by the values calculated from other walls with similar amounts of concrete, a panel system wall with 6 inches of concrete and R-20 insulation should have a mass-corrected R of close to 30 in the warmest U.S. states, and about 22 in the coldest ones.

Thermal mass

The only thermal mass inside the inner layer of insulation in the stay-in-place panel system is the sheetrock. Thus it's probably not particularly suitable for passive solar or load-leveling applications, and we didn't find it used for any.

Water resistance

Although EPS foam is a water-resistant material (as shown by the vapor permeance data in chapter 4), the manufacturers are careful not to claim that the foam in their panels provides a water or vapor barrier. Perhaps this is because there are many breaks in the foam from the ties and various fasteners. Instead they recommend damp-proofing below grade and covering all foam above grade with siding. Our builders followed these recommendations, except for a few who didn't bother to use damp-proofing. No one, however, reported water leakage through their walls to us.

Maintenance

The necessary maintenance on the panel system walls is pretty much determined by what siding you use. You'll paint more often for a wood siding (recommended for every 4 years) than a PC stucco (6 years) or a precolored PB stucco (more than 6 years). With a stucco, since it's going over a flexible material, you want to be sure of the material and the quality of the work. Northern builders using panel systems usually opted for a PB stucco and fiberglass mesh (which is really an EIFS without the additional foam) because its greater flex is supposed to help it weather the climate extremes with less chance of cracking.

Manufacturers caution that the UV rays of the sun will degrade EPS over time. This isn't usually much of a problem because the foam in the panel systems is covered with siding on the above-grade floors and damp-proofing and soil below. But the warning suggests that you shouldn't leave your foam out for long periods or leave any exposed on the final wall.

Some builders have also expressed concerns about insect damage to the outside layer of foam. The manufacturers claim that foam has no food value to insects, so they have no incentive to eat it. But there are rumors that they'll sometimes nest in it, reducing the R-value of the wall and weakening the foam that backs the siding. We couldn't confirm an actual instance of this occurring, but the rumors are widespread enough that many builders felt it was worthwhile to take precautions. The most commonly used measure was to install a termite shield (a layer of sheet metal over the foam at ground level).

Disaster resistance

The manufacturers said that panel system walls can generally be expected to have the disaster resistance of a conventional reinforced poured concrete wall. They haven't had their final walls separately fire tested, but if these walls perform like a conventional poured wall, the 6-inch version should achieve a rating of at least two hours, after which the cool side of the wall might exceed the allowed temperature limits. Reinforced concrete walls rarely crack or fail structurally in the fire test.

How well the walls resist wind and earthquake depends on how much re-bar you choose to put in them. R-FORMS' manual shows a detail of an 8-foot version of their wall with only 4 inches of concrete, one vertical #5 rebar every 12 feet on-center (o.c.), and one horizontal #5 bar at the top of the wall that they claim to be rated at a wind speed of 130 mph.

The manufacturers haven't published claims for seismic resistance, but because the walls are completely poured, you can go up to high levels of strength for only the additional cost of the rebar and the labor to place it. The cost for one #4 bar every 4 feet vertically and horizontally is about $.25 per square foot. If you wanted to double it to 2 feet by 2 feet, that would add about another $.25.

Product availability

Both UC Industries (the manufacturer of R-FORMS) and Lite-Form can deliver their product anywhere in the country within a week. But you have to pay shipping charges. These are usually about $1.25 per mile for the entire load. How far you have to ship will depend on your distance from the nearest distributor. Both companies have distributors around the country, although they tend to be closer together in the Midwest and Southeast. So if you live in the Northeast or West, you run a greater chance of having to ship a couple of hundred miles.

For R-FORMS, you need to buy the ties and, if you don't already have them, the special bracing brackets that you can reuse on later jobs. You can buy your own foam off the shelf from your local supply house, or order it from the R-FORMS distributor with the rest of the materials. For Lite-Form you need ties, but there are no special bracing brackets. You do need the specially cut and notched foam planks, however. You either buy them from Lite-Form with the ties, or buy a special notching tool (about $700) from them and cut and notch off-the-shelf panels yourself.

Labor availability

According to our builders, there are few crews experienced with stay-in-place panel systems, so they opted to train their own. Their experience indicates that the best crews have either an experienced concrete finisher or an experienced carpenter or, better still, both. These trades have the skills necessary to install straight and true walls and bracing, install door and window bucks, and deal with concrete problems like blowouts if they occur.

Because the forms are light and the erection process is systematic, a crew of three to five seemed to work best for our builders. The experienced form workers liked that they could go home at night relatively clean (usually no form oil is necessary because the forms aren't removed later) and not physically exhausted from heavy lifting .

Some electricians and plumbers were initially skeptical about working on foam panels. Accordingly, they tried to bid up the price of their work. But if the builder went through each of the tasks and showed the subcontractor how it was done, the sub usually realized it wouldn't be difficult and bid the same as for conventional construction. One or two of the builders we spoke with said that after the experience of a few homes they were able to get their subs to reduce the cost of their work marginally. This was because of the ease with which boxes and chases were cut into EPS.

Building department approval

The manufacturers say that they're really just selling a new forming system for a poured concrete wall, so building departments should treat their walls just like a conventional poured wall. Most of the departments we heard about did just that, and this made things relatively easy for the builder. The rules for poured walls are usually well established, and so long as you follow them, you generally don't need to have an engineer analyze your plans.

But if you're in a town where the officials are more cautious, you could have to go to extra lengths. There are no evaluation reports from the model codes to inform and reassure the building departments about the structural characteristics of the final wall. But if you have to go to an engineer, he or she is likely to recognize the similarity to conventional walls and be able to analyze the plans quickly.

Some inspectors were concerned that they couldn't see the concrete after the wall was poured because the foam covered it. This made it difficult to be sure that there were no voids. Builders told us that they suggested the inspectors check for voids by stabbing an awl or ice pick into the foam wherever they wanted, and that this satisfied them.

Required calendar time

Once they have experience with it, builders say they can build as fast with panels as with frame. One Midwest builder said that it took him seven days with a four-man crew from start of excavation to roof truss installation on an 1800-square-foot ranch house with a basement. The crews went home less tired than they did with conventional construction, and cleanup was faster.

Remember that there's more time spent in advance planning with panel systems than frame. It might add a little to your schedule if it slows you from getting started, but everyone agreed it was well worth it. It's a lot easier than making corrections after the concrete is poured.

The panel systems can go up in a wide range of weather conditions. Although you shouldn't pour in heavy rain, it's all right for the foam to be wet so long as you've drained the standing water out of the cavities. And rain after the pour is little problem, especially if you've covered the top of the walls. Subfreezing weather during or after the pour is often acceptable because the forms insulate the concrete well enough to keep it from freezing before it cures. Some builders told us that they've poured in temperatures as low as an amazing zero degrees Fahrenheit, even if there's no accelerator in the concrete. Dry weather is also little hindrance because the forms keep moisture in.

Crew coordination

The panel systems eliminate some of the work of the traditional trades and change the timing of some of the rest. The forming/foundation contractor can be eliminated altogether. Because your trained panel crew usually has concrete expertise, it can do the footings and then use the system for the basement walls.

The framing contractor no longer builds the exterior walls. However, the connections of the conventionally framed parts of the house (interior walls, floors, windows, doors, and the roof) to the exterior walls must be preplanned accurately. This usually means getting the panel crew and the framers to agree on the connectors, pockets, bucks, and nailers that will be pre-installed in the wall so that they'll be locked into place and available for the framers when and where they're needed.

To get sleeves set in the exterior walls before the pour, some builders asked their electrical, plumbing, and HVAC contractors what penetrations they would need in advance, then relayed the information to the panel crew. Others had the trades install their own sleeves just before the pour.

The interior work of the electrical and plumbing contractors is done in new ways (often cutting into the foam to make channels), but its timing doesn't change. These two subs come after the panel crew is finished and don't generally need special provisions to be made for their interior lines. The work of the insulation contractor is reduced to the attic space.

Connecting to wood

Our builders used three different methods to connect floor decks to the exterior walls: a shelf, a ledger, and pockets. And one builder suggested a fourth method, "bolt blocks." With the shelf method, the floor literally sits on the ledge formed when the concrete of a lower wall is thicker than the wall above it. It's easy to build a deck on such a shelf, but it's not always practical. To get a 3-inch shelf (usually considered the minimum to get enough bearing for conventional joists), you might need to pour a lower wall that's thicker (and thus more expensive) than you planned.

According to our builders, the ledger board was the method of choice. You use a series of preplaced, horizontally laid J bolts to attach the ledger tight to the concrete after it has set, and then attach the joists to the ledger with hangers. We heard one builder speculate that you could use hurricane straps instead of J bolts, but so far as we know, this hasn't been field tested. Some manufacturers say that you can leave the foam on the wall where the ledger attaches to it. But most of the builders preferred to cut away the foam so the ledger was in direct contact with the concrete. They were concerned that if the foam were left between the ledger and the concrete, it would compress over time and cause the floor to sag.

Builders who used joist pockets formed them by cutting a slot into the panel every 12 inches or 16 inches at deck height and sliding a pre-oiled block of wood into each slot before the pour. After the pour but before the concrete had completely hardened, a laborer knocked out the blocks to open up pockets in the concrete. Most codes require that the joists in the pockets be fire cut. Because the joists will be in contact with the concrete, the code might also require that the joists be pressure-treated lumber. We don't know if every building department will accept this, but one builder we spoke with used ordinary kiln-dried lumber and wrapped the tails of the joists with garbage bags so they didn't come in direct contact with concrete.

The proposed "bolt block" method is actually another way to attach a ledger. A square section of the foam of, say, 10-inch-x-10-inch is cut out where the "J" bolts are to be placed, and this is replaced with a block of 2-x lumber that just fills the cutout. The bolt is placed through a hole drilled in each block. After the pour, the blocks are locked into place and a ledger is attached to the bolts. This saves cutting the foam after the pour, requires less lumber to be in contact with the concrete (which can save money if you're restricted to using pressure-treated lumber next to concrete), and cuts out less of the insulation. However, so far as we know, it has yet to be field tested.

Our builders attached the roof either with hurricane straps embedded in the concrete or with a top plate attached to the top of the wall with J-bolts or straps. The specifics of these methods, and the pluses and minuses of each, are as discussed in chapter 4.

To attach interior partition walls, most builders tried to line them up with the ties in the exterior walls and screw the partitions into the tie heads. If the design didn't allow for this, then most builders simply nailed the partitions to the deck and ceiling members below and above. If the builders had chosen to use furring strips, they sometimes simply ran a series of extra strips horizontally and attached to those. Interior partitions can also be attached with traditional concrete fasteners driven or shot after the pour through the foam and into the concrete. One builder went to the effort of preplanning where every interior wall would fall and setting hurricane straps or J bolts into the exterior wall before the pour to attach to later.

Almost everyone mounts windows and doors on pre-installed bucks. For window bucks, they took care to make the horizontal 2-x wood (the lintel and sill of the buck) extend under and over the vertical 2-x sides. Otherwise, the uplift force put on the buck during the pour by the concrete could distort it. R-FORMS sells a special 'Y' channel (shown in Fig. 7-2) to secure the panel to the lumber of the buck. With Lite-Form you glue or nail it to the foam.

To attach interior trim and fixtures, you can screw to the tie ends. A popular alternative was to leave out rectangles of sheetrock in key spots and replace it with furring strips or sections of plywood screwed to the ties or fastened into the concrete. The wood acts as a continuous nailer. Some builders considered it more solid than the plastic tie ends.

Utility connections

Electricians cut channels for their cable and squares for their boxes out of the foam after the wall was up. Plumbers cut channels for piping. The cuts can be made with a hot knife or a router. The crews we saw were about evenly split on which tool they used. The knife appeared to make a cleaner, faster cut, but fewer tradespeople have one in their tool kit.

To attach boxes, electricians usually fastened the back directly to the concrete or used glue or a spray foam to adhere them to the foam of the panel. Using adhesive to attach to the foam was especially popular in high-volume developments because it's fast. But some builders felt it wasn't secure enough, or they thought their building departments wouldn't approve.

Some preplanning is necessary in locating the utility penetrations through the exterior. The method most often used by our builders was to insert a sleeve (usually PVC) through the block before the pour to make a channel for the lines to slide through. You need to talk with your subcontractors to figure out, in advance of the pour, the locations and sizes of openings to leave for gas, water, sewer, HVAC, cable, vent, and/or electrical lines. Another method sometimes used was to pour the concrete and then drill the necessary holes with masonry bits. You can always use drilling as a backup if you go with sleeves but forget some of them.

Change flexibility

If you make a change before the pour, it's usually less expensive than it would be with frame, according to our builders. You're moving a lightweight material that cuts easily. The work amounts to cutting foam, then taping or wiring it back together. After the pour, changes become more problematic. Builders said that design changes at that point require concrete cutting specialists, who charge a setup fee of around $200.00 and then about $15.00 per lineal foot of cutting. This works out to about $455.00 to cut the opening for a typical door. To add onto the wall involves drilling the concrete to insert rebar to extend into the addition, as well as the relatively easy job of setting and pouring the new forms.

The cutting and rebuilding might only take a fraction of the profit from their projects, but our builders still felt it was far preferable to plan ahead and do things right the first time.

Manufacturer support

The manufacturers have some prepared instructions. Lite-Form sends a video to first-time users and has concise written materials that cover some of the key points of installation. R-FORMS has a fairly detailed manual. The manufacturers also provide telephone support. You can generally get questions answered either at the company headquarters or by your local distributor or area representative. Some builders got distributors to stop by if they were close to the job site. Opinions varied on how helpful these representatives were, as some didn't build themselves. Builders who were too far to get a visit from a rep sometimes talked instead with another local builder who had used the system. The companies can usually put you in touch with experienced builders if you ask, and many of the builders we interviewed felt this was the most helpful support of all.

Other considerations

Many people remarked that walls formed of stay-in-place panel systems are quieter than frame. The manufacturers have no sound transmission class numbers for their walls. However, a conventional 6-inch concrete wall tests at about STC 48. The addition of the foam would, theoretically, increase that number, but we can't say by exactly how much.

Our builders felt, and our visits confirmed, that the building sites of stay-in-place projects were much cleaner than those of conventionally built homes. When the panels were used for the foundation, there was no form stripping, so no site revisit by the forms contractor, no stacking and restacking of forms, no whalers, and no form oil. When panels were used for the above-grade walls, there were no large piles of stud ends on the site. We observed a cleanup crew that consisted of one laborer going around the site and filling a couple of trash bags with scraps of foam.

Further information

Lite-Form, Inc.
P.O. Box 774
Sioux City, IA 51102
(800) 551-3313

R-FORMS, Inc.
10999 Prosperity Farms Rd.
Palm Beach Gardens, FL 33410
(407) 624-2515

Distribution Polycrete of Montreal, Inc.
7390 Henri Bourassa Blvd. East
Anjou QC H1E 1P2
Canada
(514) 493-6954

Stay-in-place grid systems

The grid systems get their name because that's what the concrete would look like if you were to strip the forms away. The concrete has the shape of a grid of posts and beams, something like a huge window screen or breakfast waffle. The forms are always some sort of foam block, usually with teeth or tongues and grooves that make for easy stacking and aligning. The blocks have two faces of foam that are 1–3 inches thick (one on the interior and another on the exterior) connected by either a foam web or an internal tie made of metal or plastic. Figure 7-9 shows a block molded completely of foam, webs and all, and Fig. 7-10 contains a diagram of one joined by metal ties. Figures 7-11 and 7-12 show stacked walls of all-foam and metal-tie blocks, respectively.

7-9 *Demonstration of the lightness of an all-foam block.* Featherlite

There are several brands of stay-in-place grid systems that we know to be currently available in the United States. A couple are so new that we don't have information on builders who have used them yet. Table 7-1 lists them all and some of their major characteristics.

The blocks vary in size. The smallest have the dimensions of a standard concrete block (8-inch-×-16-inch-×-8-inch). Several brands are 16-inch-×-4-feet and 10 inches thick. The larger units allow workers to erect bigger sections of wall at once, but they require more cutting.

The Polysteel Form ™
(U.S.Pat No. 4,879,855)

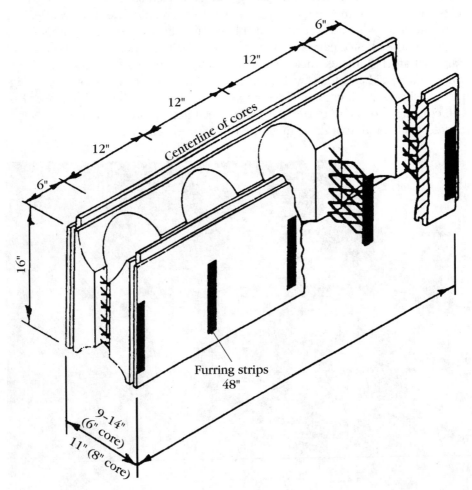

7-10 *Diagram of a block connecting the foam faces with a steel web. This web also doubles as an embedded stud.* American Polysteel

The blocks are stacked into walls, then the cavities are filled with concrete. In some systems, all vertical cavities are filled, and the webs or ties connecting the front and back are made small so that the concrete can flow horizontally as well. This forms a wall of concrete broken only every few inches by a web or tie. Figure 7-13 shows a cutaway view of this type of system. Others are designed to produce a grid of "posts" and "beams" spaced as close together or as far apart as you want. You pour only into the vertical cavities you want to fill.

7-11 *Formwork constructed out of an all-foam block.* Reddi-Form

7-12 *Formwork constructed out of a block with steel webs.*

7-13 *Cutaway view of a wall made with blocks completely filled with concrete.* Reddi-Form

The webs go up the full height of the block so they don't let concrete flow horizontally. On courses where you want a horizontal concrete "beam," you use a special block with a cut-down web called a *lintel block*. Figure 7-14 contains a cutaway diagram of walls formed with this type of system. These systems can use less concrete, although you have to be certain to fill enough of the cavities to produce a wall of adequate strength for your situation.

In several of the brands that have metal or plastic ties connecting the front and back faces of foam, the end of the tie can also take and hold a screw. The ties thus act as a sort of "embedded stud" or nailer, that's useful for connecting things to the wall before the pour (such as bracing) or after the pour (such as sheetrock).

A few systems have unique features. Energrid is made not of ordinary foam, but of a mixture of cement and EPS foam beads. This makes the blocks heavier but stronger. They require less bracing than other brands, and they can take and hold a screw anywhere across their interior and exterior faces. Yet they can be cut with ordinary wood saws. They also come in large sizes, up to 30 inches by 10 feet. The larger units put up a lot of wall area at once, but they require some heavier lifting and might produce more waste at openings and corners.

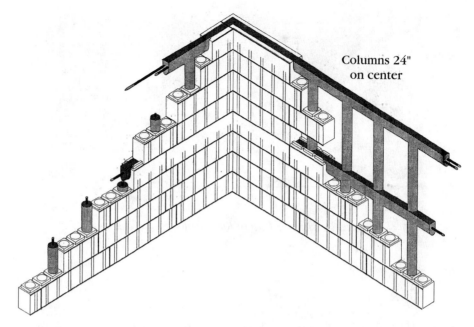

Columns 24"
on center

7-14 *Cutaway diagram of a wall made with blocks filled intermittently with concrete.* Keeva International

American ConForm has a typical product, a 10-inch-x-40-inch block with cut-down foam webs. But it also offers a version that consists of two foam faces shipped separately. The crew slides plastic ties into grooves inside the faces to form the blocks on-site. The depth of the wall can be changed by using ties of different sizes. Figure 7-15 shows the full product line.

EnerG Corp. offers a block that combines various features from other types of block. It has large blocks, but they have separate cavities so that it can be filled only in narrow posts and beams. It also has ridges inside the block with furring strips that can be screwed or nailed. This makes it possible to run pipe or cable between or through the strips and attach sheetrock and interior trim and fixtures directly. Figure 7-16 shows a diagram of a wall built of EnerG's "FFCF" block.

With almost every system, you need to use a concrete with a small aggregate size and a medium or high slump. This ensures that the mix will flow thoroughly through all the pathways inside the wall. The manufacturers give the exact concrete specs they recommend in their literature.

With most stay-in-place grid systems, the crew starts at the footing if the foundation will be made of the system. Or they start at the top of foundation if it won't be made of the system. They snap chalk lines to mark the position of the walls. Then they line up a course of blocks around the perimeter. It's important to get this first course extremely level, or the workers spend a lot of time compensating in the courses up above. Some manufacturers recommend that you lay the block in the wet concrete of the footing or foundation below while it's still soft so you can ad-

7-15 *American Conform's line of Smart Blocks.* American ConForm Industries, Inc.

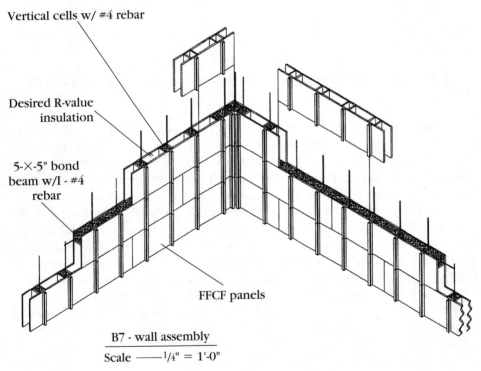

Vertical cells w/ #4 rebar

Desired R-value insulation

5-×-5" bond beam w/I - #4 rebar

FFCF panels

B7 - wall assembly

Scale ——— 1/4" = 1'-0"

7-16 *Diagram of a wall constructed with EnerG's block.* EnerG Corp.

just heights of individual blocks as necessary to get the top of the course level. But most builders we talked with found this difficult to do because the blocks tended to float out of position. They preferred to wait for the concrete below to set, then shave off the bottoms of blocks with a rasp as necessary to go over high points.

There are usually dowels (rebar sticking out of the footing or foundation) that the blocks are set over. Getting these positioned near the centers of the grid block cavities is important, so the crew might have to tug on them a little if they aren't properly aligned. To get the alignment just right, one builder we spoke with drills holes in his footing to put the dowels into after he's stacked his first course of blocks.

After the first course, most brands go up like Lego blocks. Each course is generally staggered, like conventional concrete blocks. Because most brands have teeth or tongues and grooves, they're easy to align. More and more builders are now gluing all the joints between the blocks and between the block and the footing/foundation. Some manufacturers say it isn't necessary, but the builders feel it is a small precaution to take to reduce the chances of a blowout or the forms shifting during the pour. The "glue" they use is usually one of the expanding foam sealants that comes in a can. The experienced builders bought the commercial variety that attaches a special gun (sort of like a caulking gun) to the can. With this type, there's more foam per can and the deposit of the material is easier to control.

Some brands of grid systems have blocks that form 90-degree corners without cutting. For others, you make corners by the miter method. Cut two blocks at a 45-degree angle and glue or wire them together.

Some manufacturers state that no additional bracing is necessary, but most builders used a little. Many attached an occasional 2-x-4 brace to the top of the wall and the bucks. Then they could adjust the braces to plumb the wall just before the pour. They also often put strapping over wall sections that might be prone to a blowout. When the block used had a screwing surface these straps were usually screwed directly to the block.

Vertical rebar usually gets set into place conventionally, wired to the dowels and the horizontal rebar to keep it straight. This is easy to do as you work your way up the wall. Most blocks have built-in cradles where you can set the horizontal rebar without any wiring or extra steps, as seen in Figs. 7-11 and 7-12. Reinforcing levels vary with local codes and circumstances, but typical patterns are one #4 bar every 2 feet or 4 feet, both vertically and horizontally, plus some extra around openings.

Pours are generally done in lifts of about 4 feet. Some experienced builders said they don't have blowouts, but they were all prepared for them anyway. Some just watched for a bulge, then jammed and braced squares of plywood up against them before they could break. The wall would be uneven in this spot, but the foam could be sanded flush later. More common was the method used with the panel systems: the crew cleaned the concrete out of the blown section, replaced the blown foam, and sandwiched the wall between two squares of plywood clamped together with threaded rod that ran through the wall and through predrilled holes in the plywood. If you use a system with a screwing surface, you don't even need the threaded rod. You just screw two squares of plywood directly into the studs (one over the blown section and one on the opposite side to guard against any back pressure). When we saw a blowout on a grid system house, this whole repair operation took less than ten minutes. Figures 7-17a and 7-17b show the key steps.

7-17a *Blowout that has been emptied and cut into a clean square.*

7-17b *Blowout that is being braced by attaching plywood to the embedded studs in the block.*

The heaviest concentrations of distributors and home building using the grid systems are in the Southwest, the Southeast, the Midwest, and the Plains States. However, every area of the continental United States has quick access to at least one brand, and most have nearby distributors of at least two or three.

Stay-in-place grid homes have to date appealed mostly to buyers from the quality, energy, and safety market segments. Their attraction is flexibility of architectural features and finish details, consistently high R-values, and the strength of reinforced concrete, all at a premium of about $2000 over the cost of frame. Sales have grown steadily across the country for several years now.

Labor and materials' costs

The builder quotes we got suggested that a typical 1200-square-foot ranch house on slab should cost about an extra $1300 to $1800 with one of these systems, compared with 2-x-4 construction. This comparison assumes the exterior finish is stucco in both cases. Other finishes can make the cost difference a little greater because some of them require special prep work on the grid forms.

Table 7-3 presents unit costs that were in the middle of the quotes we got from builders. The blocks themselves average about $2.00 per square foot, delivered. But this can vary widely. Keeva, one of the simplest, all-foam blocks, can be bought for $1.50/sf at the factory. I.C.E. Block, one of the blocks with an

Table 7-3. Representative Costs for Exterior Walls of Stay-in-Place Grid Systems1

Item		Cost per Square Foot
Structure and Insulation		
Blocks		$2.00
Concrete		.80
Rebar		.25
Glue		.10
Lumber		.45
Labor		.70
Subtotal, structure and insulation		$4.30
Exterior siding and finish, materials and labor		$2.00–4.00
PC stucco and paint	$2.00	
PB stucco	4.00	
Clapboard, with furring strips and paint	3.70	
Interior finish		$1.00–1.50
Sheetrock and paint	$1.50	
PC stucco and paint	1.00	
Total, including finishes		$7.30–9.30

[1] The costs listed here are in the middle of the range of quotes we received from builders reporting their actual project costs.

embedded stud, costs about $2.15. Shipping costs add virtually nothing if you're close to the plant, up to about another $.20/sf if you're a thousand miles away.

A typical 6-inch-core block takes about 0.014 of a cubic yard of concrete for each square foot of surface area. This works out to about $.80/sf if you pay $55 a yard for concrete. But depending on your local concrete price, that can be as low as $.50 or as high as $.95. Also, with some of the blocks that let you fill only some of the cavities, the amount of concrete can be cut by over half, depending on how much strength is required for your wall. Rebar is about $0.25 per square foot if placed at 4 feet o.c. both vertically and horizontally, which is a typical spacing. Most of the 2-× lumber used for bracing can be reused. You might cut its cost of $.45/sf if you recycle it efficiently. The $.70 for labor includes stacking and gluing, cutting out openings and assembling bucks for an average number of doors and windows, wiring the corners, and some miscellaneous work (cleanup and the like).

When the exterior finish is some form of stucco, it's a little cheaper than frame because the foam surface is already in place. The stucco coats and the mesh are the only parts that have to be applied. Inside, you can save some money by applying stucco directly to the foam for a sand finish. A few builders did this.

Learning costs

The builders we spoke with felt that the costs of the exterior walls of their first couple of homes might be up to 10 percent more than the walls of houses they built after they got their use of the system down. This works out to an extra $1000–2000 for a 1200–2000-square-foot ranch house. They also said that by their third project they and their subs felt as comfortable with their grid system as with conventional construction. One builder in Maine told us he couldn't see ever going back to frame.

The builders themselves had to spend extra time on the first couple of houses in planning for block placement around corners and openings, planning bracing of walls, and general coordination and supervision. The most important task of the supervision was pushing the crews to make sure the footing and the forms were true and level.

Even if they're experienced with concrete, the forms crew has a few things to learn with grid systems. The stacked blocks are not as forgiving as traditional forms, which can be racked while still allowing for a level pour. So the crew needs to start on a footing or foundation that's "dead level." And they have to spend the time to make sure that the erected foam walls are true before the pour. During the pour, there's so much pressure on the blocks they might try to shift or "walk." This means the bracing must be accurate and adjustable, so corrections can be made during the pour. It's very expensive to rip down or shim a nontrue wall once the concrete has set. It was also necessary for the wall crew to learn how to handle blowouts. But once they did, they agreed that it was faster and easier than with conventional forms.

The plumbing and electrical trades were initially skeptical. But if the sleeves for penetrations through the exterior walls and the interior bolts and fasteners had been properly installed, all seemed to go as well as with a conventionally built home. The builder has to keep after the trades to make sure they specify and lay out all of these things before the pour. The only exception is that when using a system that has thick foam webs or is filled in only some of its cavities, you can often put sleeves through later with little difficulty by cutting them through a spot where the wall is only foam. Our builders also told us that it's important to follow the manufacturers' concrete specs. If the manufacturer says to use a 5-inch slump and 3000-psi concrete (which is typical), they do so.

Buyer reaction

In the Southeast and Southwest, where homes built of some type of concrete are common, our builders indicated that grid homes were received about the same as any other concrete home. In their marketing, the southern builders stressed the overall energy costs (both heating and cooling), wind resistance, and reduction in fire and wind insurance costs. They generally sold to quality segment customers, but with their low concrete costs and volume production methods, some southern builders sold successfully to the value segment.

The homes we heard about in the rest of the country were not spec built, as in the South, but custom. The customers chose the grid systems because of a combination of their reduction of heating costs, resistance to moisture and pests, lower insurance costs, flexibility of design, and their inside quiet. The customers outside of the South were usually typical quality segment buyers. However, some fit the description of the energy or safety buyer.

Exterior and interior finishes

Stucco is the finish of choice above grade, partly because it's the easiest to install. It's usually applied by scratching the foam surface with heavy sandpaper or a wire brush, then applying a base coat and a top coat. For PC stucco, some grid system manufacturers recommend installing a mesh before the first coat. To install it, some builders suggested that before the pour you push wire lath or pieces of single wire through the block to act as tie wires for the mesh. The concrete holds them firmly after it's poured. When PB stucco is used, it's usually part of an EIFS with its own fiberglass mesh. No separate mesh is used, and the fiberglass is not anchored into the concrete.

If you use a system with screwing surfaces, you can also screw vinyl or aluminum siding directly to the wall. For nailed sidings like clapboard or shingles, you first screw furring strips to the studs on the outside, at a cost of about $.20 per square foot. The builders' experience was that screwing clapboard and shingles directly will split them. If the system doesn't have a screwing surface, you can attach furring by embedding some type of connector in the concrete before the pour. One popular trick was to push 4-inch double-headed nails from inside the block through the exterior face. After the pour, the nails are held by the concrete, the furring is pounded onto their protruding points, and you can bend over the excess point to give an even better hold. Another method was to push in pieces of wire or metal straps or ties at predetermined points to secure the furring until after the pour.

Below grade, most manufacturers recommend damp-proofing, and most builders used it. You have to be careful what you use, however. Some petroleum-based products eat away the foam. A latex-based emulsion was the most popular choice. Most good building supply houses carry them.

For interior finishes, we were told about a number of different alternatives. Some builders doing affordable housing stuccoed the interior directly, just as they did the outside. The builders who attached sheetrock did it by the same methods used on stay-in-place panel systems.

Design flexibility

The builders we spoke to used the grid system either below or above grade or both. One builder simply used the product below grade, then marketed the idea of finishing it into a warm and dry basement for relatively little increased cost to the buyer, but high profit to the builder. Others used the product below grade and only went the minimum required by code (from 1 foot in the South to 4 feet in the North). Above-grade, single-story houses were most common. But several builders went up two stories, and several manufacturers say that three or four are no problem.

Corners and bays are generally made by cutting the blocks at an angle and wiring or taping them together at the joints. For 90-degree angles, some manufacturers now offer special corner block or have notches built into their regular block. Irregular openings can usually be cut out of the wall in-place. There's an example in Fig. 7-18.

7-18 *Cutout for a circular window ready for the remainder of the foam to be removed.*

Because most corners and openings are cut by hand, the more of them you have, the more your costs will rise, just as with frame. One builder we spoke with estimated that for each additional corner (over four) he simply added one man day of labor. He figured that additional charge would cover the increased costs of handling the product, cutting, extra bracing, wiring and/or taping the joint, and waste.

Curved walls could be made in two different ways. For large, gradual curves, some blocks can simply be bent slightly and held in place with bracing. Once they're poured, the concrete holds them permanently. Sharper curves are made by cutting each block in one section of the wall at an angle and gluing or taping them back together. This forms a series of slight angles that come close to a true curve. Figure 7-19 shows a striking example of the formation of curves.

When you're using stucco, the foam surface also gives a lot of flexibility in exterior trim. To add raised features, you simply nail or glue on pieces of foam in the desired shape and stucco over them.

7-19 *Forms for a high-end residence with curved walls and irregular corners.*
American Conform Industries, Inc.

R-value and energy efficiency

Table 7-1 gives the manufacturers' stated R-values for their systems. They're almost all over R-20 because of the two layers of foam. Most of the calculations use the insulation R-value method. While this is the simplest method, it might be a little more accurate for these systems than it is for some others. The insulation is usually continuous, with little or nothing in the system cutting through it. The insulation breaks in the final wall occur mostly on the inside layer, where you can cut out the foam for utility lines and floor decks and interior walls and fixtures. The exterior foam stays largely intact.

The completed walls have thermal mass of approximately 7–12 Btu/sf/degree, depending on the thickness of the core. This is more than enough to receive extra insulation credit under most energy codes. The systems that fill only some of the cavities with concrete will be proportionately lower in thermal mass, however. But almost all of these systems have a higher R-value than is required by U.S. codes, so the credit isn't really necessary. Most companies did not calculate a mass-corrected R-value for their walls. But a few with continuous concrete in their final walls did. They got values of about R-30 for southern climates.

Thermal mass

Because most of these systems have a layer of insulation between the concrete and the interior, their interior thermal mass is technically zero. We didn't find them used in any passive solar or load-leveling homes. However, Energrid's block contains some cement that provides thermal mass adjacent to the indoor air.

Water resistance

Although molded EPS is a relatively water resistant material, the manufacturers and builders we talked with didn't rely on the foam alone to keep the envelope dry. Nearly all of the manufacturers suggest that their products be waterproofed below grade with compatible (nonpetroleum-based) waterproofing products. Some, such as Reddi-Form, even give you a list of compatible products and their distributors. All of the manufacturers also require that above grade their blocks be covered with a conventional exterior siding. Nonetheless, it's interesting to remember that as a backup you have a layer of the same material used to make disposable coffee cups. We got no reports of water leaking through the walls into one of the grid homes.

Maintenance

Painting is determined by the siding used, about every four years if you use a wooden material on furring strips, six years if you use a cement-based product like PC stucco, and less often if you used a vinyl or aluminum product or a pre-colored PB stucco.

When we asked our builders about the longevity of stucco applied to foam, they said that they had few callbacks because of problems in that area. They did say that there might be cracking where the sunny side of the house meets the shady side and suggested that a PM or PB mix would be more durable. Experience with stucco over foam in commercial buildings confirms the value of polymer stuccoes wherever the wall will be subject to sharp differences in temperature.

Some manufacturers point out that EPS degrades under UV rays. They suggest that the blocks not be directly exposed to sunlight for any length of time. This is usually not an issue because the foundation walls are almost always damp-proofed and covered with soil, and above grade they get an exterior finish. As with the panel systems, some builders installed a termite shield over their block at ground level as a precaution against insects burrowing in the foam.

Disaster resistance

The manufacturers that have put their completed walls through the fire wall test generally find that they survive for about two hours, after which the cool side of the wall exceeds its maximum allowable temperature. But note that we have no test results from the systems that get filled in only some of their cavities. Some builders expressed the opinion that these walls would not last as long because the fire could melt through their large sections of empty foam.

Because grid walls are poured all the way around, you can add rebar to get them up to various levels of strength for wind and earthquake resistance at relatively little incremental cost. Some of the manufacturers have engineering calculations to show the amount of rebar needed for different resistances. For example, for ICE Block with a 6-inch core, using one vertical #4 bar every 2 feet produces a wall with a wind speed rating of just over 160 mph. Other manufacturers claim that their blocks achieve Zone 4 seismic ratings with high levels of rebar. We heard of some very cautious builders in California who installed one bar every foot to achieve high earthquake resistance. Because the details change with the brand, you have to consult the manufacturer on this.

Product availability

Availability varies widely by product. Some brands are made in several plants and are sold by distributors around the country. At the extreme, the manufacturer of Smart Block (American ConForm) claims to have a network of almost 300 distributors with 15 field support staff willing to travel to your site and give assistance. Some other brands are made in one plant and sold only direct from the factory. However, in almost any area of the country there are at least two manufacturers with established distribution. The situation is changing fast as new brands emerge and sign up distributors in more locations. You simply have to call the manufacturers of the brands that interest you to find out where you can buy them.

Almost every company will ship anywhere for the cost of the truck. This amounts to $1.00–1.50 per mile for the entire load. In practice, builders rarely order from a plant that's more than a few hundred miles away because shipping costs add up.

Delivery time also varies. American ConForm claims that in most locations their product is immediately available from the distributor, although some of the builders we spoke with indicated that delivery time was more like a week to ten days. Other manufacturers, like Keeva and Energrid, indicate that there can be lead times of up to 45 days to get their products. The builders agreed that it pays to order in advance and get your distributor to review your design to make sure you're ordering enough to account for waste.

Labor availability

Builders' success rate at finding experienced people was low. There are not yet many workers experienced with these systems, and most of the builders didn't want to commit to hiring someone with specialized skills when they couldn't guarantee a continuous stream of work.

So most of the builders we spoke with trained their own crews. They said that the skills they needed for an effective crew were those of a good framing carpenter and those of a good concrete finisher. Our builders either took their in-house labor with those skill sets, or they went out and recruited them. The crews we observed were usually four or five people. The crew chief was either the experienced carpenter or concrete finisher, and the second member of the crew picked up the remaining skill, with the other two or three members being laborers. All the builders said that their jobs had too many workers, and they believed that an experienced three-person crew could do the job efficiently.

Initially the electrical and plumbing subcontractors were skeptical about working on EPS block walls. They were especially uncertain about estimating how much time the work would take them. But after working with the products, their only additional work seemed to be in the up-front planning of exterior wall penetrations, requiring a properly located sleeve.

Building department approval

As listed in Table 7-1, many of the brands have evaluation reports from one or more of the model code bodies. Builders using them reported that their local building department was usually satisfied with the report, if it was from the code body covering their region of the country. However, occasionally a very strict department might also require an engineer's stamp on the plans. For the brands without a report, you can expect departments to ask for engineering more often. Some of the manufacturers will send additional documentation to the inspectors and talk with them by phone if you ask. We never heard of any grid system being turned down for a permit so long as the builder complied with the building department's request for documents and plans.

Sometimes the inspectors became concerned about the electrical work. Usually they feared that the wiring wouldn't be deep enough behind the interior face of the wall, or the boxes wouldn't be securely attached. There's enough foam in almost all of the systems to bury the cable about an inch and a half (including the thickness of the sheetrock), which is more than enough for the majority of codes. The inspectors usually are satisfied once you show them this. They're also usually happy if you attach the boxes to the embedded studs. If your system doesn't have them, the inspectors might be satisfied if you simply attach the boxes to the foam with glue or a spray foam. Others prefer that you use fasteners through the back of the box to attach to the concrete.

Required calendar time

From the builders we learned that after their first or second project, the time required to construct a wall was the same or less than that of a conventionally framed house, with less site cleanup. One builder in Ohio said that he could have a 3200-square-foot (living space), two-story house with roof trusses erected in seven days. But planning was key. If you're the type of builder who pushes the daily production and figures that all errors can be corrected in the finish, a word of caution. It's not easy to correct mistakes once the concrete has been poured. Our builders felt that extra time spent in layout, leveling footings, squaring walls, and educating the crew and the subs was well spent. The old adage has never been truer, "Measure twice, cut once."

As with the stay-in-place panel systems, the use of the foam forms can extend your construction calendar if you're in a cold climate. Because of their water resistance, pouring in hot or dry climates is also easier as there's less need to take extra steps to make sure the concrete cures properly.

Crew coordination

The special steps necessary for coordinating crews are almost the same ones necessary with the stay-in-place panel systems. The wall crew does a lot of work that a separate forms crew and the framing crew would do on a frame house. Most builders take care to plan for any penetrations or fasteners that need to go in the exterior walls so they can insert sleeves or the fasteners before the pour instead of drilling them into the concrete later. This includes planning some of the connection details between the exterior wall and the framing (how floor decks and roofs will attach, for example) in advance. They usually involved the framers in this. Because the crews can carve channels in the foam with a hot knife or router, no pre-planning is necessary for the inside runs of electrical and small plumbing lines.

Making connections

Most of the key connections to grid walls were made the same way they were to panel system walls (see the previous section). The key difference is that the grid blocks don't have plastic ties to fasten to. Some have embedded studs or furring strips that can be screwed to instead, one (Energrid) has a continuous screwing surface, and some have no screwing surface, so you have to resort to other options for connecting to the wall.

Our builders used the same three methods to connect floor decks to the exterior walls that they did with panel systems: a shelf, a ledger, and pockets. The details were exactly as previously described under stay-in-place panels. Figure 7-20 contains a photo of joists pocketed into a grid wall. They attached the roof either with hurricane straps or a J-bolted top plate, as with panels.

7-20 *Floor trusses pocketed into a stay-in-place block wall.* American Polysteel

Interior partitions were also attached in the same ways as with stay-in-place panels. The builders who used blocks with an embedded stud or furring strip could screw directly to the studs or strips if they lined up with the point where the interior wall intersected. If the interior partition fell between studs, they sometimes ran strapping horizontally and attached the partition wall to it. Those using a block without screwing surfaces used one of the other measures previously mentioned under panels. Those using Energrid could screw to any point in the block.

Windows and doors nearly always went on bucks. The bucks were usually set on the block wall as soon as it reached sill height, and then the crew continued to build the wall around it. Some builders glued it to the block afterward to hold it securely for the pour, and, of course, it was braced with extra 2-× to hold the force of the concrete until the pour hardened.

To attach interior trim and fixtures, you could again screw to embedded studs (if your system has them) or cut away the foam and attach a wooden nailer to the concrete.

Utility connections

Electricians and plumbers cut out the foam to fit in their lines and boxes just as they did with the panel systems. Figure 7-21 shows a completed cutting operation for an electrical box.

To attach boxes, electricians usually screwed them to the embedded studs, if there were any. If there weren't, they used one of the two methods used with the stay-in-place panels: fastening to the concrete or gluing to the foam.

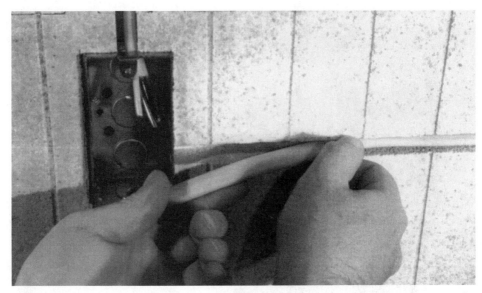

7-21 *The installation of electrical wire into a channel cut into a stay-in-place* *wall.* American ConForm Industries, Inc.

Builders using systems that have metal webs and that get completely filled with concrete planned out where they needed utility penetrations so they could insert PVC sleeves before the pour. An example is in Fig. 7-22. If instead you use blocks that have foam webs, you can cut narrow sleeves through after the pour with ease. Just go through one of the webs. With systems that don't fill all the cavities, you can cut out even large sleeves later through an unfilled part of the wall. You need only cut through foam.

Change flexibility

Changing the wall is another matter that's about the same for the grid systems as it is for the panel systems. Changes are quite easy before the pour. They consist of cutting foam and taping or wiring it back together. But they're quite difficult and expensive to alter after the pour. This requires concrete cutting and adding on to the concrete and steel reinforcing. The one additional consideration with the grid systems is that they aren't of constant thickness on the inside (the way the panels are). So when you add on to an existing wall, whether it's before or after the pour, you sometimes need to cut the new blocks at a specific point so that the thickness of the new and the old blocks where they connect is the same.

Manufacturer support

The manufacturers' support is very dependent on the product and how close you're located to their plant or distributor. Some manufacturers will send a local field representative out to your site to help you get started if you're nearby. Some will also send a rep to distant jobs, but they often require that you cover travel and expenses. Most of the manufacturers also offer support by tele-

7-22 *PVC sleeve installed in a stay-in-place block wall to accommodate a dryer vent.*

phone, and feedback suggests that it can help clear up a lot of basic questions. Some of the more long-established companies also have detailed manuals. But the favorite method of getting help was to talk with other builders that have used the product before and get their advice. The manufacturers could often help by giving the names of local builders who had used their system.

Other considerations

The grid systems that have had their completed walls sound tested report STCs of about 48 without finishes. Depending on what finishes you use, this will go up by 2–3 points in the final wall.

The job sites of grid houses were as neat as those of stay-in-place panel houses. There were fewer lumber cutoffs than with frame or conventional forms, and there was little of the oil or other accessories of conventional form work. Cleanup consisted mainly of walking around the site and tossing stray bits of foam into a plastic trash bag. Especially when they were inexperienced, crews sometimes scattered concrete on the floor during the pour. This someone would have to chip apart and clean up.

AAB Building Systems
840 Division St.
Cobourg, Ontario Canada K9A 4G9
(905) 373-0004

American ConForm Industries
1820 South Santa Fe St.
Santa Ana, CA 92705
(800) CON-FORM

American Polysteel Forms
5150 F Edith NE
Albuquerque, NM 87107
(800) 977-3676

EnerG Corp.
4203 West Adams
Phoenix, AZ 85009
(602) 470-223

Energrid, Inc.
6421 Box Springs Blvd.
Riverside, CA 92507
(602) 386-2232

Featherlite Building Products
301 West Hills Rd.
New Canaan, CT 06840
(203) 966-2252

GREENBLOCK Worldwide Corp.
PO Box 749
Woodland Park, CO 80866
(408) 625-1898

I.E. C. Block
570 South Dayton-Lake View Rd.
New Carlisle, OH 45344
(800) 423-2557

Keeva International, Inc.
1854 North Acacia St.
Mesa, AZ 85213
(602) 827-9894

Reddi-Form, Inc.
593 Ramapo Valley Rd.
Oakland, NJ 07436
(800) 334-4303

On the horizon

Permaform of Ridgefield, Connecticut has recently introduced a somewhat differ-ent stay-in-place forming system. Instead of foam, the system is made of PVC plas-tic. The basic unit is octagonal and interlocks with its neighbors. Widths of 4–8 inches are available. We're told it's been used to construct thousands of units abroad; the first few houses in the United States are now nearing completion.

Western Forms of Kansas City has developed a modern reusable forming system that builders are beginning to use for the above-grade walls of homes (although it's already been used widely to build single-family homes in other countries). The aluminum forms are light enough for a single worker to lift, and they never rot or rust. Their greater-than-conventional width (3 ft) permits a greater labor efficiency in form setting, which the manuacturer estimates at 30%. Many size variations and special finish patterns are also available. Some home builders have used the system to make their entire exterior wall structure out of reinforced concrete, finishing with conventional materials inside and out. The manufacturer is currently marketing it to the production builder market, where crews could use the forms especially economically by constructing repeated units without many changes in the form setup.

Futher information

Permaform U.S.A., Inc.
7 Doubleday Lane
Ridgefield, CT 06877
(203) 438-8357

Western Forms, Incorporated
6200 Equitable Road
Kansas City, MO 64120
(800) 821-3870

8

Shotcrete systems

Shotcrete is concrete sprayed from a hose and nozzle under pressure from a pump. Using shotcrete to form walls is new to U.S. home building, but it's been used in other countries for years. In the United States, we're more accustomed to shotcrete as a material to coat steel bridges and form grain silos and swimming pools.

There are two varieties of shotcrete: "wet" and "dry" (or more precisely, "wet mix" or "dry mix"). Wet shotcrete consists of cement, sand, and sometimes pea gravel premixed with water before the pump pushes it into the hose. Dry shotcrete, also called *gunite*, isn't mixed until it reaches the nozzle. The pump pushes a dry mix of cement, lime, and sand down one tube of the hose, and it pushes water down the other tube. These mix just as they're spraying.

Houses are built of shotcrete in the basic way silos are. Workers erect a frame of wire mesh or rebar in the shape of the exterior walls, then spray it. The concrete covers the steel and hardens to form a reinforced wall.

A few innovations have made this process suitable for home construction. One is the incorporation of insulation into the mesh or rebar. The others are a series of new tricks for installation that make it possible to run utilities inside the wall and get an attractive finish with little extra difficulty.

There is currently one shotcrete system that's actively sold for residential construction in the United States: 3-D from Insteel Corporation. But there are some other systems whose manufacturers are ready to take orders.

3-D

The basic building unit of the 3-D system is a 4-foot-wide panel of steel wire mesh on either side of a foam core. Crews connect these panels to form a frame for the walls, then apply shotcrete over the mesh to complete the structure. The result is a continuously reinforced concrete sandwich wall of tremendous strength with good insulation in the center. In addition, it's relatively easy to make irregular angles, curves, and openings because the lightweight panel cuts and bends quickly.

As we might expect with features like these, the buyers to date have been typical of the quality market segment. They have wanted homes that are at least a little better than conventional construction in almost every respect, and they have been willing to pay more to get them. But particular interest has arisen among residents of disaster-prone areas. They're lured by the high, unbroken strength of the system, which promises a phenomenal ability to withstand wind and seismic forces. Habitat for Humanity has also built several affordable housing projects with Insteel. However, their projects are typically done with donated labor and materials, not on a for-profit basis.

Insteel Construction Systems, Inc., the manufacturer of 3-D wall panels, ships product from Georgia and California to builders throughout the Southeast and Southwest. But the company will send them just about anywhere for the cost of shipping, and Insteel has long-range plans to open plants in other parts of the country. To date, about 70 percent of the houses built in the United States are in the Southeast.

Figure 8-1 shows a diagram of a 3-D wall panel. The mesh on each side of the panel is made of wires welded on 2-inch centers. It can be ordered with bright (ungalvanized) or galvanized wire. In every square foot, nine additional galvanized wires pierce the insulation. These "truss" wires are set at alternating angles and are welded to both layers of mesh to act as a structural bridge between the front and back.

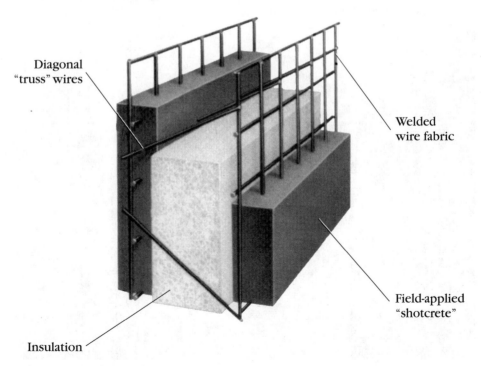

Diagonal "truss" wires

Welded wire fabric

Field-applied "shotcrete"

Insulation

8-1 *Side view diagram of a 3-D wall panel.* Insteel Construction Systems, Inc.

The panels come in several thicknesses. One size that's commonly used in homebuilding is 4 inches, measured from one face of wire mesh to the other. The foam is 2½ inches thick, and the final, shotcreted wall is 5½ inches. Some builders use a thicker, 5½-inch panel with 4 inches of foam. It produces a final wall about 7 inches thick. Any panel can come with a mesh of either 14- or 11-gauge wire. All our builders chose the heavier 11-gauge wire. Recently Insteel also began offering panels with 12½-gauge, which is a little less expensive than the 11. The panels can also be made in any length, although the builders we interviewed used 4-foot-×-8-foot panels for most of their work.

The wall erection crew sets panels over dowels protruding 12 to 18 inches from the foundation, as in Fig. 8-2. The dowels fit between the mesh on one side and the foam. The workers tie adjacent panels with wires by hand or with a pneumatic tool that costs about $600. Insteel sells the tool, but you can get a similar model from a lot of suppliers. However, you don't simply tie the panels directly to one another. A separate piece of mesh, called "cover mesh" is laid over the joint for strength and continuity and tied to each panel, as seen in Fig. 8-3. The cover mesh and an angled version ("corner mesh") that connects panels joined at an angle are also available from Insteel.

Wooden braces, visible in Figs. 8-5 and 8-7, hold the panel walls upright. To provide resistance to the force of the flying concrete, the braces typically go opposite the side that will be shotcreted first. The crew adjusts them in advance of the shotcrete to get the panels plumb.

8-2 *Setting 3-D wall panels on a foundation.* Insteel Construction Systems, Inc.

8-3 *First floor of 3-D wall panels erected, and reinforced at the seams and corners with cover and corner mesh.* Insteel Construction Systems, Inc.

8-4 *Cover mesh reinforcing at the corners of window openings.* Insteel Construction Systems, Inc.

Openings can be cut before or after the panels are erected. However, every builder we spoke with did it after. You get to cut at eye level, and there's little chance for error. You can use a reciprocating saw with a metal cutting blade for a fast, one-step cut, or you can snip the mesh with bolt clippers and use a keyhole or drywall saw to rip the foam afterward. Cover mesh, set at an angle, gets placed at the corners of an opening, as in Fig. 8-4. This provides resistance to diagonal cracking at those points.

A sort of wooden border goes around all openings and on top of the wall. The border is visible in Fig. 8-5. To form it, you cut the foam 1¾ inches farther back from the edge of the opening, then insert a 2-x-4 or 2-x-6 (depending on the thickness of the panel you're using) in place of the foam between the two faces of mesh. You staple the mesh on to lock it in, then nail a 2-x-6 or 2-x-8 to the side of it. In a door or window opening, this second, inner layer of lumber forms a buck. On top of the wall, it forms a top plate. It's always made of 2-x lumber that's deeper than the panel and centered so that it protrudes ¾ of an inch on either side. That way it will serve as a guide for the depth of the shotcreting and line up with the final wall surface. All of this lumber is pressure-treated.

For windows, you can also set a prefabricated window frame in place of the inner layer of lumber. A third option is to place U-shaped mesh around the jambs, lintel, and sill to form them out of reinforced concrete. You then bolt a masonry-type window with no flanges directly into the concrete jambs. But the builders we spoke with generally used the bucks.

8-5 *View of window openings after fitting with 2-x bucks.* Insteel Construction Systems, Inc.

8-6 *Shotcreting in progress.*
Insteel Construction Systems, Inc.

When the panels are set, the crews place the reinforcing and rough in the utilities. Because the mesh serves as continuous reinforcing, extra rebar goes only at stress points. Some builders put no extra steel over narrow (less than 4 feet) openings. Over larger ones, almost all of them placed two #3 rebar, one inside and one out. Over very long openings, they put a couple of bars on each side, one above the other. Some builders also added one bar vertically at each corner, although the manufacturer says this isn't necessary. Rebar fits neatly between the mesh and insulation. Where you need extra room, you can melt a groove in the insulation with a heat gun or torch.

The shotcrete goes over all exposed mesh. Figure 8-6 shows the job in process. It's applied in two applications—one "structural" layer and one "finish" layer. Figure 8-7 shows a wall structure with both layers on and troweled. Trailing the nozzle man are workers with screed boards who flatten the applied concrete. They press their boards against the 2-x bucks and top plates, using them as depth guides. According to the manufacturer, some crews also stretch "piano wires" in front of the panels, spaced to the intended finish depth. They serve as depth guides. Separate finishers follow with trowels to further smooth the surface or texture it, depending on what the final finish will be. Figure 8-8 contains a cutaway of a final concrete-steel-foam sandwich.

8-7 *Wall of 3-D after finishing the shotcrete.* Insteel Construction Systems, Inc.

8-8 *Cutaway of a panel covered with screeded shotcrete.* Insteel Construction Systems, Inc.

Labor and materials' cost

The cost quotes we got were all from the Southeast. Builders of average-sized (about 2000 square feet of living space) homes said their overall costs with 3-D ran about $5,000 higher than frame. One builder of very large, expensive homes (around 3000 square feet of living space and selling prices of $350,000–500,000) quoted a cost difference of about $10,000 per house. In all cases, the builders used the thicker (5½-inch) 3-D panel and were comparing their costs to 2-x-6 construction with a high-quality stucco or wood siding.

Table 8-1 shows costs per square foot that were in the middle of the quotes we got. Some quotes were as much as $1.50 lower to $1.00 higher than the range shown in the table, depending on several factors. A big one was whether the labor was local. Some builders brought in out-of-town crews that had more experience with the product and paid them travel plus a per diem. Another factor is the height of the structure. Higher-up scaffolding is necessary for panel erection, shotcreting, and concrete finishing. We saw quotes that estimated the labor on the structure roughly doubling after the first 10 feet. Most of the houses used to derive the figures in Table 8-1 were 12–20 feet, so they averaged the low costs of low construction and the high costs of higher walls. If you go up only one story, your costs might be lower than the range of the table.

Most of our quotes also included shipping costs of a few hundred miles (included in the materials in our table). If your project is nearer the plant, you might save a few dimes per square foot for shipping. Our structure quotes don't include a subcontractors' markup, however. If you hire a sub to handle the structure and insulation, instead of ordering the materials and supervising the work yourself, you could pay 10–30 percent more for that part of the work.

Learning costs

All the builders felt the system was very easy for them to learn. Usually they read the manufacturers' materials and visited a job to watch 3-D going up, then they felt they knew all they personally needed to know about the mechanics for their first job. The greater learning costs came in finding satisfactory crews and (for the trades) developing the skills to do the shotcreting efficiently.

Some builders went through one or two shotcrete crews before they found one that met their standards. This took time and sometimes meant that on the first house or two, an extra crew went back to the site to correct something the builder didn't like—usually the finish. In addition, both the panel erection and shotcrete crew worked more slowly on their first jobs. Between both of these factors, the labor bill on the first house might be 50 percent higher than the sixth or seventh. This works out to around $3000 extra for a typical 2000-square-foot (living space) home. The panel erection crew typically got near their full efficiency after only two houses. The shotcrete crew might take three or four.

Table 8-1. Representative Costs for 3-D Exterior Walls[1]

Item		Cost per Square Foot
Panels (including shipping)		$1.90–2.15
4-inch with EPS insulation	$1.90	
6-inch with EPS insulation	2.15	
Cover mesh, wire ties, and corner bead		.25
Concrete		1.00
Lumber for openings		.25
Panel erection labor		2.25
Shotcrete and finishing labor		1.05–1.25
Rough finish, both sides	$1.05	
Smooth finish, both sides	1.25	
Subtotal, structure and insulation		$6.70–7.15
Exterior siding and finish, materials and labor		$ .80–2.50
Elastomeric paint only	$ .80	
PC stucco and paint	1.00	
PB stucco (colored)	2.50	
Interior finish, materials and labor		$ .80–1.50
Paint only	.80	
Sheetrock, furring, and paint	1.50	
Total, including finishes		$8.30–11.15

[1]The costs listed here are in the middle of the range of quotes we received from builders reporting their actual project costs.

We heard about a couple of problems the shotcrete crews had to work out. One was getting the proper concrete mix and shooting it so that it sticks to the mesh correctly. They told us a 0- to 2-inch slump does the trick, with the exact formula determined by the way the particular crew works and its equipment. The manufacturer recommends a 3000-pound mix with coarse sand and no gravel. When the crew decided on the best mix, they also had to do some experimenting to find a local ready-mix company that would deliver a consistent batch of it every time. And finally they needed to figure out who the finish people would be and how they'd coordinate with the rest of the crew.

Buyer reaction

Some builders have reported a strong favorable reaction to 3-D from very high-end home buyers. One who specializes in that market offers the system as an option to all of his customers and finds that about 10–15 percent choose it over frame, even though he charges substantially more for it. He reports that they respond to the insulation, durability, resistance to wind storms, and lower insurance costs. He claims, "This is our Cadillac, and they want that."

A few buyers we heard about keyed in on the system's disaster resistance. One couple relocating from California to the Carolinas "had been through earthquakes and fires and knew about our hurricanes, and they saw this concrete and steel and had to have it. The house was only half done, and they saw it Friday morning and signed Friday afternoon."

Exterior and interior finishes

Most of the houses we heard about had a separate coat of stucco as the exterior finish. Because it went onto concrete, the cost quotes we got were lower than for stucco on frame: about $.50 per square foot for PC stucco (unpainted), and $2.50 for PB with coloring.

If the concrete gets the right finish, it can be painted directly. The builders we spoke with that did this specified concrete with no coarse aggregate and had the crew trowel it smooth for a sand finish. They also used elastomeric paint, although as far as we know, latex will do. They preferred the premium water protection because they were building for an upscale market.

Logically, any other sidings should go on the same way they go onto other concrete walls. Attached furring strips (about $.20 per square foot for materials and labor) could accommodate screwed and nailed sidings, and a ledge on the foundation would support a brick veneer.

Inside, all our builders used sheetrock. Some glued it directly to the walls and held it in place with a few concrete nails, while some attached it to interior furring strips. Some crews told us they put a sand finish on the inside without sheetrock, but this was only in commercial buildings.

Design flexibility

Although we never heard of 3-D used as a basement, the manufacturer says that it can be. They recommend that an engineer review basement plans because some extra rebar might be required.

The system seems capable of doing just about everything else. We've seen it go up 20 feet, and the manufacturer says it's been used on single stories 36 feet tall. Odd wall angles are simple to make because the panels cut readily. Curves can be formed by cutting several short sections of panel, forming them into a series of small angles, attaching cover mesh at the joints, and putting a true curve on this frame with a skilled troweler smoothing the shotcrete. One builder formed curves out of a single panel without cover mesh, although the manufacturer doesn't recommend it. He clipped a strip of, say, an inch wide out of the mesh face on one side, then cut out a pie-shaped strip in the foam that also ran the length of the panel. After cutting out such a "wedge" in the mesh and foam, he could bend the panel a few degrees and tie it back together on the cut side.

With such a cut every several inches, the panel bent into a series of angles that served as the frame for a curve.

To make irregular openings, you simply cut the shape you want out of the erected panels. Rebar goes above. You then improvise something to form the wooden buck that goes along the edge.

R-value and energy efficiency

The manufacturer sells the panel with several different thicknesses of foam and a variety of different types. Most of our builders opted for a 5½-inch (before shotcrete) panel with a 4-inch sheet of foam. They also chose molded EPS. The R-value of the foam itself in such a panel is 18. The 4-inch panel holds a sheet of foam 2½ inches thick. If that's also molded EPS, the foam is R-11. If instead you pay an extra $.50–.75 extra per square foot for extruded polystyrene, you can get R-values for the foam of 20 in the 5½-inch panel.

The one regular break in the insulation is the "truss" wire that connects the two faces of mesh. Insteel hasn't yet calculated or tested the R-value of a final wall, complete with a layer of concrete on either side and truss wires connecting them. The insulation might also be broken or reduced where it has been cut out or melted back for rebar or utility lines. But these are usually a small percentage of the total wall area, and, in practice, few of them are complete breaks.

The thermal mass of the wall will vary depending on how thick the shotcrete is. But with the normal thickness of 1½ inch on each side, there will be about 30 pounds of concrete per square foot, with a thermal mass of about 7 Btu/sf/degree. This is enough to qualify for an R-value credit under most energy codes. Our builders didn't apply for the credit because the customers they sold to were interested in getting higher insulation anyway. But if you wanted to use the thinner panel and its R-value didn't quite meet your code, you might get over the top by figuring in the correction for thermal mass.

Thermal mass

About half of the mass of a 3-D wall, or 3–4 Btu/sf/degree, is inside the insulation envelope. A couple of buildings in the West built of 3-D were designed to take advantage of this for passive solar heating.

Water resistance

The manufacturer and the builders we spoke with favored a polymer-based coating on the outside of the Insteel wall, whether it was a good coat of latex paint, an elastomeric paint, or PB stucco. The wall itself has a continuous surface of a dense form of concrete. It rarely cracks, and behind the concrete exterior is a layer of foam broken only by the truss wires. However, the builders preferred the extra protection to be sure.

We found no complaints of any water getting through the walls of a 3-D house. They're all fairly new (the oldest were 4 years), but this is a positive sign, nonetheless.

Maintenance required

From the reports we got, maintenance on this wall system beyond painting is virtually zero. If the concrete is properly cured and cover mesh goes at the corners of openings as recommended, cracking is almost nonexistent. The manufacturer explains that there's too much steel too close to the surface for significant cracking. The builders say they haven't had to do any repairs yet.

You get proper curing if the concrete remains wet for about a week after it's applied. Our builders sent someone by once or twice a day (depending on how dry the weather was) to wet the walls down. According to the manufacturer, you can instead spray a "curing compound" (a thin film that holds moisture in) over the wall surface.

If you paint the shotcrete directly, repainting probably needs to be as frequent as with any concrete surface—about every six years, according to paint manufacturers' recommendations.

Disaster resistance

The overall disaster resistance of 3-D is striking. Walls covered with 1 inch of concrete on each side have a fire rating calculated by engineers at 1½ hours. If you put 2 inches of concrete on each side, the fire rating increases to 2 hours. According to the manufacturer, at very high heats the insulation can melt, but it won't support combustion.

Builders along coastal Georgia and the Carolinas are meeting local wind codes that require a 120-mph rating using only panels with the thinner (14-gauge) wire mesh. When 11-gauge wire is used, the houses are frequently rated at 150–200 mph. Habitat for Humanity built a house out of 3-D in Homestead, Florida about a year before Hurricane Andrew. Photographs taken afterward show it, structure intact, surrounded by other houses that were leveled.

Four buildings in California made of 3-D survived the Landers earthquakes of 1992. The buildings were part of a research facility in the Mojave Desert. They were mostly glass on one side, and the tallest went up 24 feet, so they had a lot of unsupported wall area. The Landers earthquakes consisted of two shocks, one of Richter 6.5 and one of 6.9, with epicenters about 50 and 75 miles from the site. According to an engineer who examined the buildings afterward for Insteel, "There was no sign of any crack or damage of any kind to superstructures and foundations."

Product availability

Insteel manufactures its materials in Brunswick, Georgia, and Mexicali, Mexico. For U.S. customers, it ships from Brunswick and Calexico, California (on the

Mexican border). It will ship a truckload (which holds up to enough to do 8000 square feet of wall) almost anywhere for about $1.20 per mile. Insteel is considering other plants, but has no definite locations or dates as of the time we're writing this.

Other than the panels and accessories that Insteel supplies, no unusual products are necessary. The builders and crews generally ordered a week in advance. Delivery was usually within six to eight days to distant locations. And, they told us, it is reliable.

Labor availability

Building with 3-D requires a crew to erect the panels and make the bucks and top plates, and a shotcrete crew that also finishes. Work with the panels is by all accounts easy to learn. A decent carpenter and some solid laborers can do it. The carpenter is necessary to make sure the bucks, bracing, and leveling are done correctly.

The shotcrete crew is a little trickier because there are, traditionally, few crews with exactly the right combination of skills. Structural shotcrete crews who do things like bridges and silos usually know how to get the concrete over the mesh correctly, but they don't have fine finish people. Gunite crews who do swimming pools usually have a decent finish person, but they're less familiar with the structural side of the work.

There are now a few shotcrete crews in the Southeast who specialize in 3-D. A couple of high-end home builders we spoke with simply brought in one of these crews and paid for travel plus a per diem. It worked well, but, of course, it was more expensive. Some other builders used a more local structural shotcrete crew and lined up plasterers to follow them to do the finish. This had mixed results. The plasterers were usually unfamiliar with the relatively dry concrete and didn't always get the desired finish. In addition, many shotcrete crews normally use a mix containing pea gravel. If someone didn't tell them to leave out the gravel, the finish had streaks. But perhaps with experience the builder and the crews could work out the difficulties.

Building department approval

Insteel has a complete evaluation report from the National Evaluation Service (No. NER-454), which testifies that the 3-D system can meet the requirements of BOCA, ICBO, and SBCCI. In addition, the company frequently sends representatives to the local building department to acquaint officials with the system before a builder's plans hit their desk.

The builders we spoke with said they never had any difficulty getting approval. On some large projects, they got an engineer's stamp on the plans. But for this they went to Insteel, who retains an engineer to design details and stamp at a reasonable rate. The builders also said there was nothing special the inspector had to do in inspections. They all went pretty smoothly.

Required calendar time

The builders calculated that forming the walls of a house with 3-D takes a little longer than with frame. For the large (3000–5000 square feet of living space) houses our builders specialized in, quotes for exterior wall construction time ranged from two to three weeks. This involved crews of about five erecting the panels and another four to six doing shotcrete. They felt that erecting the walls would have been shorter with frame by about one day per week.

The panel erection took about one-third of the total time, and shotcreting took two-thirds. It's possible to start shotcrete before all the panels are up, so you can shave some time off the calendar by overlapping crews. The panels can go up in most weather. Shotcrete is normally done above freezing, although northern crews might have tricks to go slightly below. Like most other types of construction, it can't go on in hard rain. Hot or dry weather is no problem so long as the concrete is properly cured.

Crew coordination

Builders agreed that it's no harder to coordinate crews with 3-D than with frame. The sequence is similar. One crew erects the walls, the utilities trades cut into or through them as necessary to install their lines, the walls are covered, and then the utility crews come back for finish work. The major difference is that the shotcrete crew is inserted in the middle. After the rough-in, they shoot both sides of the wall, and the siding people (if any) and sheetrockers follow.

A couple of steps before and after shotcreting ensure a clean job. The crew usually puts tape over electrical boxes and plumbing outlets beforehand. Someone needs to take that off while the concrete is still wet, or crack it off with a hammer after drying. The bucks and top plates need to be cleaned with a wire brush wherever there's excessive concrete buildup so they'll still take nails later on. And loose shotcrete is cleaned off the slab (if the slab is down at that point). You can put plastic sheet or sand on the slab ahead of time to make this easier.

The total number of crews can increase by two: the panel setters and shotcreters. But that goes down by one if your panel crew and your framers are the same, and it goes down again if you use the concrete itself as the exterior finish (because there's no siding crew).

Making connections

For intermediate floor decks, the preferred method of connection was to attach a ledger to the wall with steel bolts. The bolt goes through the foam so that it's anchored into both the exterior and interior concrete. Roof framing goes directly on the top plate installed by the panel crew. In high-wind areas, the panel crew notched the top plate and put hurricane straps through it to secure the rafters or trusses above. The bottom of the strap they wired to the inside mesh. Shotcreting locked them into the wall.

Doors and windows went directly on the bucks installed by the panel crew. If the interior was furred, interior fixtures and trim were attached to the furring. If not, the finish carpenter used one of the concrete fasteners for most objects. One crew working on an unfurred wall screwed a half-inch strip of pressure-treated plywood to the base of the walls for nailing baseboards. The favorite way to attach interior walls was to cut nail a pressure-treated 2-x to the concrete and nail the wall section to it.

Utilities installation

If the walls are furred out inside, small lines go between the furring strips. If they aren't, the electrician and plumber can run lines under the mesh surface.

When the panels are up, the crews can slide conduit or pipe in the space between the mesh and foam. If the space between isn't large enough, they can make channels with a heat gun or blow torch. Either one melts a groove in the foam. According to Insteel, you can also use a can of spray paint. It eats a groove in the insulation and marks the layout clearly. Everyone we spoke with who ran wiring inside the panel put it in plastic conduit. It appears to be required by most codes, and it's a good practice in case people want to repair or reroute the wiring later. The mesh has to be cut out to make space for electrical boxes, as shown in Fig. 8-9. You can make room for very large lines like vents by melting out the foam completely along a column. But this eliminates insulation on that column and is therefore not a good idea for water pipes.

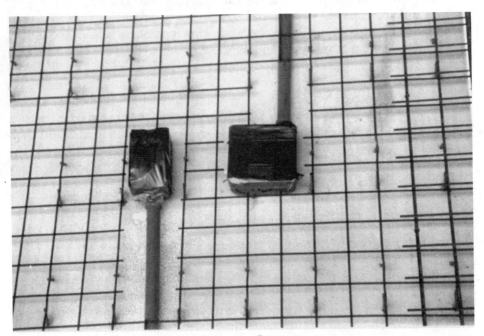

8-9 *Electrical boxes and cables installed in 3-D wall panels.* Insteel Construction Systems, Inc.

Shotcreting covers the lines and locks everything into place. Boxes and plumbing outlets are covered with duct tape ahead of time, and the tape is removed later.

Change flexibility

Before shotcrete, changes are easy. Cuts are made with the same tools used to cut the openings in the first place. Filling in is done by cutting pieces of panel and patching them on with cover mesh. Moving a window sideways might take one worker a couple of hours.

Some builders did a thorough check of the panels after they were up to spot any possible problems before shotcreting. One even took the client through to make sure everything was the way he wanted it.

After shotcreting, changes are "no fun." The ones we heard about started by chipping the concrete with a pneumatic chisel down to the wire mesh to expose it, then snipping the wire with bolt cutters. According to the manufacturer, it's also possible to cut all the way through with a diamond-bladed saw. Moving a window could take a day and cost several hundred dollars, although most of our builders didn't have hard figures on this. They simply planned ahead and avoided late changes.

Manufacturer support

By all accounts, Insteel is eager to make sure builders have plenty of support when they use the company's new product. There are instruction manuals that cover the major steps of construction in clear language. The company tries to set new builders up to see jobs in progress nearby so they can learn by watching a real building go up. We've heard of houses near their plant for which they contributed their own people to help set the panels. This saved the builder a few dollars and made sure the job was done right.

For builders struggling with plans, Insteel offers the services of an engineer they have lined up who can rework frame plans and stamp the final design. You pay for this, but the engineer knows the system and can work efficiently.

The company also sometimes visits building departments to educate them about the system before they get plans. We never heard of any 3-D plans getting rejected by a building official.

Other considerations

Several people remarked that Insteel houses are quiet inside. In testing, they achieved a sound transmission class of 35 without sheetrock and 40 with sheetrock on one side.

As in any shotcrete operation, 5–10 percent of the concrete comes back as "rebound." That is, it doesn't stick to the wall, but bounces back a bit and falls to the ground. Outside, this wasn't considered a problem. It's just scraped up or broken and worked into the soil. On the inside, you can just leave it if you shotcrete before pouring the slab. When builders shotcreted after the slab or floor was down, most swept or scraped up the rebound before it could dry and used it as fill. To facilitate this, the manufacturer suggests putting sand or plastic sheet over the slab before shotcreting.

Further information

Insteel Construction Systems, Inc.
2610 Sidney Lanier Drive
Brunswick, GA 31525
(912)-264-3772

On the horizon

There are three other shotcrete systems that are or might soon be available for U.S. home building. Two of the systems use panels similar to 3-D's. The Therml-IMPAC panel has been used in thousands of homes over a period of more than a decade in other parts of the world. In Mexico it's known by the name Covintec. Figure 8-10 contains a photo. The manufacturer, IMPAC International, also has a plant in California, devoted mostly to commercial sales. However, the company will sell to residential projects and is considering actively marketing to homebuilders.

Truss Panels are manufactured in Panama and widely used in Central and South America. The manufacturer, Truss Panel Systems, S.A., will ship to the United States and is considering setting up manufacturing here. Figure 8-11 shows one of the panels.

Corotherm, an Alabama manufacturer of panel systems for commercial construction, has developed a shotcrete system for residential buildings so recently that it doesn't have an official name yet. The company's working name is "Corotherm Spray Wall." The system is based on panels that have a continuous sheet of oriented structure board on the inside, 6 inches of foam on the outside, and a top and bottom plate. Figure 8-12 shows one story of the panels set on a slab. The outside of the foam has grooves that contain rebar, and you can cut away foam and put in more bar as necessary on-site. A shotcrete crew covers the outside. The result is a reinforced concrete wall with thick insulation inside and a continuous interior nailing surface. Corotherm has just started promoting the system.

Further information

IMPAC International (Therml-IMPAC)
16641 Orange Way
Fontana, CA 92335
(909)-829-3089
(800)-227-9591

Steve Black (Truss Panel)
U.S. Representative, Truss Panel Systems
Estate Pastory #7, St. John
U.S. Virgin Islands 00830
(809)-776-6237

Corotherm Building Systems, Inc. (Corotherm Spray Wall)
285 Mill Road
Madison, AL 35758
(205)-772-3416

8-10 *A cutaway section of a Therml-IMPAC panel covered with shotcrete.*
IMPAC International

8-11 *A Truss Panel.* Truss Panels Systems, S.A.

8-12 *Shotcreting Corotherm panels erected to form the first floor walls of a house.*
Corotherm Building Systems, Inc.

9

Panelized systems

The design of almost any house structure can be divided into several flat or nearly flat walls. By building these wall panels as separate units out of concrete and then setting them in place, it's possible to erect a nearly seamless structure of exceptional strength in a short time. Panelized construction has been used for years to form the walls of commercial buildings. But more recently, systems designed for home construction have appeared that take advantage of a series of new advances. Streamlined reinforcing schemes keep materials costs low. Plastic foams provide high insulation. High-density concretes provide a measure of water resistance. And exterior surfaces that can take almost any siding allow the houses to have aesthetics that are similar to those of any other type of house.

In traditional panelized construction, the panels are sometimes made on-site by building forms that are laid flat on the ground and filled with concrete. After curing, the panels are tilted into place. This method is sometimes called *tilt-up*. Tilt-up panelized construction allows the forms crew to make adjustments on-site up to the moment the concrete is poured. It also allows the concrete to be brought in on a ready-mix truck, which is relatively inexpensive and convenient.

The alternative is to assemble the panels in a factory. This method is sometimes called precast panelized construction. Factory assembly sharply reduces the amount of work that has to be done in the field. It can go on in any weather and produces a consistent product. A truck delivers the finished panels. All that's left for the site crew is to set the panels and attach them to the other structural members (foundation, floor decks, and roof).

We discovered two panelized systems currently used regularly by U.S. builders: the Royall Wall System and Superior Wall. Both are precast. But there are also some very new systems, both precast and tilt-up, "on the horizon."

The Royall Wall System

The patented Royall Wall System is currently available in Southeast Florida. However, the manufacturer (Royall Wall Systems, Inc.) is currently working to license it to manufacturers and to builders who want to make their own panels in other areas as well. In Florida the system has attracted a rapidly growing number of customers because it provides high energy efficiency and wind resistance for a cost that's almost dead even with frame and conventional block in their area. At least some buyers from almost every market segment have chosen this system over the alternatives. Builders like it because it cuts their production schedule by one to two weeks and replaces three to five separate crews with one.

The walls themselves are panels with a layer of steel-reinforced concrete on the outside and another layer of molded EPS foam on the inside. Inside there are also steel studs every two feet on center, plus one running horizontally along the top and one along the bottom. These are visible in Fig. 9-1. The studs are standard "C" profile studs. Their side flanges pass through the foam and are embedded into the concrete. This makes them lie flush with the foam surface on the interior and anchors them firmly to the panel. Although you can't see it from the outside, the concrete is deeper behind the studs. This forms vertical columns (behind the vertical studs) and a beam and bottom plate (behind the two horizontals). These columns and beams of deeper concrete contain steel rebar and provide much of the structural strength of the panel.

The panels are 5½ inches thick, matching the width of 2-x-6 dimensional lumber. Royall Wall Systems, Inc. ordinarily makes them in any height up to 14 feet, 8 inches. Units larger than that can be made, but require special shipping measures. However, to build taller structures, the shorter panels can be stacked. The

9-1 *Royall Wall Systems crew setting a panel in place.*

9-2 *Installation crew attaching screw jack braces to hold up and plumb a panel.*

panels could be made any length, but they rarely go longer than about 30 feet because they would be hard to transport and set. The standard panel is designed for above-grade exterior walls. However, the manufacturer also offers a special version for interior fire partition walls used between units in multifamily construction.

The manufacturer handles design, production, and installation. Builders bring their house plans to Royall Wall Systems, Inc., which converts them to a set of panel designs. There's one panel for each straight exterior wall, except that very long walls are sometimes broken into multiple panels. After the panels are complete, the company sends them to the site on a truck, and a crane and company installation crew set them in place.

The panels are generally set on a slab on grade (the dominant form of foundation in Florida). The slab must have a 5½-inch shelf around its perimeter on which the panels go. This is visible in Fig. 9-2. The shelf serves as a guide to the set crew and helps prevent water from entering through the joint between panel and slab later on. The crew adjusts the panels to plumb and holds them in position with adjustable screw jacks used as braces, as in Fig. 9-2. The panels have shiplap joints on their side edges so that abutting units overlap a couple of inches. They're attached to each other with bolts that go through preformed

9-3 *Adjacent panels connected at top with a bolt through steel plates with pre-formed holes.*

holes along the panel edges, as seen in Fig. 9-3, and sometimes with metal plates, as in Fig. 9-4. Figure 9-5 shows a site with nearly all the panels set.

As one of its final acts, the installation crew also places grout along the inside of the joint between slab and panel to seal it against air and water infiltration.

After the set is done, it's the responsibility of the builder to seal the joints between adjacent panels on the outside. There are a lot of ways to do it and, in the words of the general manager of the manufacturer, "The builders are inventing new ones all the time." Some popular methods are to fill it with stucco and cover that with a fiberglass mesh, or squeeze in a preformed plastic gasket and cover it with stucco or a sealant textured like stucco.

Labor and materials' cost

The Florida builders we spoke with agreed that houses built with Royall Wall cost virtually the same as houses out of concrete block or 2-x-4 frame. In all cases they were assuming that the exterior siding was PC stucco (by far the most common siding in South Florida). They also assumed that the block they compared to was furred inside and fitted with ¾-inch fiberglass insulation between the furring strips (standard in their area). They assumed the frame contained R-11 fiberglass insulation (also standard). One builder said, "Royall is probably a tad more than frame on average, but it's so close that on any one job the advantage might go one way or the other."

9-4 *Panels connected at a corner both with bolts (visible along the right edge of the front panel) and a steel plate (being fastened at top).*

9-5 *A nearly completed Royall Wall installation.*

Table 9-1. Representative Costs for
Exterior Walls Constructed with the Royall Wall System[1]

Item		Cost per Square Foot
Royall Wall bill (includes engineering, panels, setting, and sealing)		$4.00
Subtotal, structure and insulation		$4.00
Exterior siding and finish, materials and labor		$ .40–.80
Paint only	$ .40	
PC stucco and paint	.80	
Interior sheetrock and paint		1.30
Total, including finishes		$5.70–6.10

[1]The costs listed here are in the middle of the range of quotes we received from builders reporting their actual project costs.

Representative unit costs for Royall Wall are in Table 9-1. Payment to Royall Wall Systems, Inc. for conversion of the plans to panel designs, production of the panels, and setting the panels totals about $4.00 per square foot of gross wall area. This can vary, but not much. If the panels are short (less than 10 feet) the standard charge is $3.95 per square foot. If they're over 10 feet, it's $4.10. There's no extra charge for openings. However, if there are large openings, the manufacturer charges less. Depending on the size of the opening, the total bill drops by an amount equal to about $1.30 to $2.00 times the square footage of the opening. There's also normally no extra charge for extra corners, non-90-degree corners, or irregular opening shapes. The costs of the various finishes that the Florida builders we interviewed used are as listed in the Table 9-1.

Learning costs

Builders claimed that there were almost no start-up costs with the Royall Wall System. This was mostly because the manufacturer and its crew take care of nearly all the unconventional tasks.

Builders using the system for the first time familiarized themselves with the basics by viewing a video and reading a brief brochure provided by the manufacturer and, sometimes, watching the construction of another Royall Wall house nearby. Occasionally they also met with the company to ask questions. They all felt that this was plenty of preparation.

The two trades that had to change their practices because of the use of the system were the rough carpenters and electricians. And even these didn't have to do much different. The carpenters had to fasten various components (roof trusses, interior walls, and so on) to the steel studs in the panels. This required them to use self-tapping screws and a screw gun instead of nails. The electricians had to run their wires and install boxes in the foam. But most of the channels for this were precut by the manufacturer, so the electrician simply had to pull the cables through. The builders claim that, so far as they know, neither of these trades charged them any more for the same job on a Royall Wall house than they would on a frame or conventional block house.

Buyer reaction

Builders told us they got a strong positive response from potential buyers by showing them pictures and cross-cut sections of panels that revealed the insides of the Royall Wall. Buyers liked the high insulation of the panel and its apparent strength. According to one builder, "Most people really preferred it. If I lost a sale it was because I didn't have the lots the buyers wanted or the designs that suited them best, not because they didn't like the Royall Wall." Another said, "Around here concrete block is normally considered the best form of construction and frame is second. But when I got a chance to show this to people, they ranked it on top."

The houses have sold to buyers in the value, quality, energy, and safety segments. The big attractions are the energy efficiency and strength (especially in wind). In addition, the manufacturer produces irregular angles and openings at no extra charge, making striking arches and other architectural features easy to include. So the buyers looking for premium features are impressed. But since all these features are available in the standard wall at a cost about equal to the cost of other types of construction, middle-income buyers (the value segment) have frequently chosen Royall Wall as well.

Exterior and interior finishes

The exterior concrete of the panels comes with a slightly rough finish to provide good adhesion by stucco. Figure 9-6 shows this standard surface.

An estimated 90 percent of the builders using Royall Wall do stucco and then paint over the panels. The manufacturer is careful not to claim that the standard surface can serve as a final finish, but we heard that a few builders have used it that way. Some of them left it exposed everywhere, simply sealing the joints and painting. Others might stucco some parts of the walls. Another finish used occasionally was to put a brick veneer over one or more walls. To do this, the concrete crew built the slab a little larger to form a brick shelf. They also ordered their wall panels with brick ties embedded at regular intervals in the concrete exterior face, which the manufacturer will provide for about $.20 per square foot extra.

According to the manufacturer, you can also attach screwed or nailed sidings (clapboard, shingles, vinyl, aluminum) by first shooting furring strips onto the panels with concrete nails. However, we didn't run across any builders who had done this.

On the interior, all builders we spoke with screwed sheetrock directly to the steel studs in the panels. The process is similar to woodframe and costs the same.

9-6 *Close-up view of the factory exterior finsh on a Royall Wall panel.*

Design flexibility

Royall Walls haven't yet been used for basements, which are rare in Florida. Some of our builders stacked panels on top of one another to go up two stories. The manufacturer says that this is as high as they're designed to go and still be self-supporting. If there is a second floor, the panels must be specially designed. The top edge of the first floor panels and bottom edges of the second floor panels are shaped so that they form a shiplap joint when stacked. They're also outfitted with preformed holes that are bolted through to connect them, or sometimes they're connected by screwing steel plates to the steel studs along their edges. The manufacturer handles all of the panel design details, and the installation crew makes the actual connections between panels. Because of the greater materials and labor costs, the company charges about $.10 per square foot extra for the second-floor panels.

Royall Wall houses can be made with any irregular angle between walls. The manufacturer simply gives the ends of the panels involved the appropriate slant. Some houses have called for a few short panels with wide angles between them to form bays or octagonal floor plans. A true curve or near-curve has not been done and, according to the manufacturer, is probably not practical because it would involve many, small panels.

Openings of almost any shape are also possible, thanks to a quick and inexpensive technique that Royal Wall Systems, Inc. has developed to make molds out of foam. Arched doorways (as shown in the color photographs) and windows are frequently requested. According to the company's general manager, "The most unusual one we ever did was a window in the shape of a shamrock."

The manufacturer claims that it doesn't charge extra for houses with irregular angles, irregular openings, or even extra angles. However, the general manager suggests that it might have to charge a higher price for a house design that involves a very large number of small panels.

R-value and energy efficiency

The foam is molded, expanded polystyrene bead board. Its thickness varies from about an inch at some points on the panel to 3½ inches. Sometimes the R-value of the panel is quoted at 20. However, this is at the points with the thickest insulation, only. As calculated by the manufacturer with the parallel path method, the R is about 14. This calculation, however, does not take into account the studs, which break the insulation every two feet plus along the top and bottom, or the little bit of extra R-value added by the concrete.

Other insulation breaks are minor. They include the bolts that connect panels at a corner and any penetrations made through the walls for utility lines and the like. Although channels are cut into the foam for wiring, they don't go all the way through the foam, and the piece that's been cut out to form the channel is normally put back into it after the cable is in place.

The manufacturer hasn't calculated a mass-corrected R-value for the panels.

Recently, Royall wall has begun to offer thicker versions of its panel with more insulation. The company calculates that the version that's about 1½ inches thicker is about R20, and the version that's about 4 inches is about R30.

We estimate that the panels contain an average of about 38 pounds of concrete per square foot of net wall area (excluding openings) and have a thermal mass of nearly 8 Btu/sf/degree. This is more than enough to qualify for preferential R-value treatment under most energy codes. However, the insulation far exceeds Florida requirements without special treatment, so, not surprisingly, none of the builders we interviewed applied for it.

Florida homeowners who had moved from frame or conventional block houses to a Royall Wall house of a similar size told us that their electric bills (including air conditioning) had dropped by ⅓ to ½. One builder told us, "It's like a cooler. I used to build them, and you get that same sudden cool feeling when you walk inside."

Thermal mass

Because the concrete of the Royall Wall is outside the insulation, the only significant thermal mass inside is the sheetrock. The system is therefore not particularly suited for passive solar or load-leveling applications, and we did not find it used for any.

Water resistance

The face of the panels is made of concrete with a load-bearing strength of 6000 psi. According to the American Concrete Institute, concrete with a strength of 4000 psi or greater is "watertight." Moreover, most builders cover the material with two water-resistant layers (stucco and paint), and the others with paint or a brick veneer.

This leaves the joints as the crucial points for potential water leakage. At the joint along the bottom of the panels, the slab is stepped down to form an outside shelf, and the installation crew presses mortar into the joint on the inside. Both of these measures are designed to stop water penetration. Forming a good seal at the vertical joints is up to the builder.

The manufacturer claims that they haven't heard of any causes of leakage in the approximately 3000 standing Royall Wall houses. The builders we spoke to said they'd never had any water penetration problems in their units.

Maintenance required

The only regular maintenance required is painting the exterior. Because the siding used is usually cement-based (stucco, or the wall panel surface itself), this should occur about every six years, versus about every four with wood products.

The manufacturer speculates that it might be necessary to reseal the vertical joints between panels after ten years. But since the oldest Royall Wall houses have been standing only five years, it's difficult to know whether this is accurate.

Royall Wall Systems, Inc. gives all buyers a 10-year warranty on the panel structure at no extra charge. There are almost no claims submitted, however. The manufacturer says that during a two-year period, the repairs it made to correct structural defects in their panels cost a mere $460.

Disaster resistance

The standard panel hasn't been put through the fire test, but engineers working for Royall Wall have estimated its fire rating from the design and materials at just over an hour. The manufacturer also offers a special version of the panel that's intended for use as a partition wall between units of a multifamily building. It has more concrete and less insulation. It also extends all the way to the underside of the roof, and above the ceiling line it's solid concrete. Its fire rating is two hours. One builder felt that its extension right up to the roof was a big advantage. It met the strictest codes without the trouble of framing and sheetrocking a gable above the ceiling.

The manufacturer can put extra reinforcing into the panels during production to meet any level of wind resistance buyers ask for. They offer a version of the panel, called the "Plus" panel, with a wind speed rating the engineers calculated at 200 mph. But they've generally been asked simply to meet the local

Florida codes, which call for relatively high ratings of 110–130 mph. There's no extra charge for panels designed to perform to a 140 mph rating.

Royal Wall Systems, Inc. hasn't yet had calculations done or had the panel tested to estimate its resistance to earthquake. The company believes that if the system is offered in high seismic areas the reinforcing can be redesigned to meet local codes and needs.

Product availability

Royall Wall Systems, Inc. and its production yard are in Royal Palm Beach, Florida. The company will ship and install panels to any point within a 50-mile radius of the yard with no transportation charge. They'll ship as far as 200 miles, but beyond 50 they don't offer installation, so builders have to arrange to set the panels themselves. There's also a transportation charge for these long deliveries, which comes to $250–300 on a typical shipment of 200 miles.

Although the panels are patented, the company is willing to license the right to manufacture its panels. It suggests that interested builders and manufacturers contact the office to discuss it. A couple of large Florida and Caribbean builders have signed licensing agreements with Royall and are now producing some of their own panels. In addition, the company is in negotiations with other firms in the United States and Canada. When these conclude, the panels might be for sale in additional areas.

Labor availability

In the Palm Beach area, the Royall Wall installation crew takes care of most of the unusual tasks, so the only trades the builder had to hire with any special skills were the rough carpenters and electricians. And these had only to know how to fasten into the steel of the wall panels with self-tapping screws (a task for the carpenters) and cut cable and boxes into foam (the electricians). For the electrician, the special skill required is lower than most people imagine because the manufacturer precuts channels in the foam for any electrical lines shown on the plans. Everyone said that the instruction needed to bring a decent crew in either of these trades up to speed would take no more than a couple hours.

Actually, the builders we spoke with were able to hire mostly experienced crews and avoid training altogether. There are now enough Royall Wall houses built in the southeast Florida area that a lot of crews have worked on them already.

Building department approval

When the manufacturer converts the builder's house plans to a set of panel designs, a company engineer checks all the work and stamps the final plans. Royall Wall supplies these to the building department as well as the builder and deals with the department if there are any questions. None of the builders we spoke with had any difficulty getting approval for their Royall Wall houses.

The builders also said that they've never had to do anything special for the inspectors on their Royall Wall homes, and they've never had trouble passing inspection for the structural portion of the houses.

Required calendar time

The Royall Wall crews normally set the panels for two complete houses of 2500–3000 square feet (living space) in a day. If there are unusual delays, they might take a full day for one house. Our builders all planned on leaving the site empty for installation crew for one or two full days. When the slab and site were nearly ready, the builders arranged a date for the set with the company. Sets scheduled for the afternoon might occasionally get pushed to the next day if the one scheduled ahead of it for the morning got delayed. After the set, the builders brought the next trade (usually the carpenters to set the roof trusses) as early as the next morning.

The main causes of delays are rain and lightning. The set crews work through light rain, but they wait out heavy showers. Lightning is hazardous because it can strike the crane, so setting of new panels stops and the crane is lowered until the lightning subsides. Hot weather is no hindrance. The panels haven't been installed in very cold climates yet, but the manufacturer sees little difficulty in setting them even in subfreezing temperatures. The only obvious problems of installation in the cold are curing the mortar along the slab joint and curing any sealant put in the vertical joints. These jobs might have to wait until the weather warms.

Crew coordination

The consensus of the builders was that coordinating crews is easier with Royall Wall houses than it is with any conventional method of construction. This is because (in the Palm Beach area) the manufacturer and the installation crew take care of a lot of tasks for the exterior walls that normally fall to separate crews or the builder, including ordering materials, assembling the structure, and insulating. As one builder put it, the system reduced construction time by one to two weeks, which became one to two weeks that he didn't have to visit the site, make calls to the contractor supply house to get every needed bit of material, and translate between different crews to make sure they all had what they needed and came on site at the right time.

The builder is responsible for specifying all the electrical and plumbing runs and penetrations. Usually they do this by making sure all these things are on the plans. Royall Wall delivers the panels with the necessary channels and sleeves built-in. On the off chance that something is missed or the plans change midstream, there are ways to cut in extra channels and holes that are not too difficult.

9-7 *Roof trusses waiting to be anchored to the panel with metal straps.*

Making connections

Roofing members are fastened to the wall as shown in Fig. 9-7. The lumber rests directly on the panel. The carpenters fasten metal straps to the steel stud on top of the panel with self-tapping screws, and to the roofing lumber with nails.

For two-story houses, the manufacturer puts holes through the first-floor panels at the floor line. The carpenters put bolts through these holes to attach ledgers inside. The floor joists rest on top of the ledgers or fasten to the ledgers with joist hangers.

Interior walls were fastened to the embedded studs if the two lined up. If they didn't, builders either nailed the wall only at top and bottom, avoiding connecting to the exterior wall altogether, or cut out foam between studs to put a block flush with the foam. They attached the block at each end to the studs and nailed the interior wall to the block. According to the manufacturer, some builders instead use steel L-angle pieces to attach an interior wall to a stud a few inches away.

Around windows and door openings, the manufacturer includes a frame of 2-x-6 pressure-treated lumber right in the panel. The builders nail a buck to this and attach the window or door to the buck just as they would to a frame wall. They pointed out to us, however, that they don't count on the installed frame to be precise. Instead they shim the buck plumb and square before they nail it in. One builder went so far as to add a buck, and then nail steel L-channel around it inside and out, adjusting the L-channel to get his square and plumb opening.

Inside, most fixtures and trim can get screwed to the steel studs. The manufacturer will also pre-install blocking directly into the panels where requested.

The builders found this particularly useful for the kitchen cabinets. With blocks all the way across at 3-foot and 7-foot heights, they could nail their cabinets into a solid backing without having to hunt for a stud.

Utilities installation

For all electrical lines specified on the plans, Royall Wall Systems, Inc. puts a sleeve in the concrete beam at the top of the panel and cuts out a channel of foam down from that sleeve alongside a metal stud. Figure 9-8 shows the result. The manufacturer also replaces the foam inside the cut-out channel. When the electricians do their rough-in, they run all electrical service in the roof space and drop one cable down for each box, just as is done in most houses in Florida. But in the Royall Wall house, they remove the foam inside the channel first, drop the cable through the sleeve and down the channel, cut out a rectangle of foam for their box, screw the box sideways to the stud, then pull the cable through the box. After that they can replace the foam over the cable to maintain insulation.

Occasionally there might not be a precut channel where an electrical line is needed (the plant accidentally missed it on the plans or the plans changed). Just in case the manufacturer leaves extra sleeves in the beam at top, so the electrician can cut away a channel of foam beneath one of these to put extra electrical lines through. In addition, the steel studs don't extend quite to the top or bottom of the panel, so the electrician can cut out channels above or below them to run from a sleeve or box in one section of the panel to almost any other section.

9-8 *The sleeve and channel for an electrical cable to run down from the ceiling crawlspace. After pulling the cable, the electrician will replace the foam strip tacked up on the right in the channel.*

As one builder put it, "If there's one thing I don't like about the system, it's the constraints on plumbing. But even those aren't too tough to deal with." According to the builders, it's difficult enough to run piping and vents inside a panel that they almost never try. While in colder climates plumbing rarely goes inside exterior walls anyway, it often does in Florida. The builders dealt with the situation by rerouting all lines through interior walls or using special valving that reduced the need for vents.

If penetrations through the exterior walls are shown on the plans, the manufacturer will pre-install sleeves. If a builder wants to add some more later, it's fairly easy. You just pick a spot where the concrete is thinnest (between the studs) and drill with a concrete bit.

Change flexibility

The builders we spoke with tried to plan carefully so that no changes in the structure were necessary after the panels were made. But when they did need a change, Royall Wall Systems, Inc. handled it. The builder would call the change in to the company. Then company engineers would check to see what would have to be done to maintain the strength of the affected panels, and the company would send out a crew to do the cutting and/or fill-in work.

According to the manufacturer, the most common change is to the dimensions of a window opening. The total charge (including engineering) for changing the size of a window is usually $100–200, and the site work takes two to three hours. The largest change the company has made was to alter the appearance of the front of one of its houses by replacing three panels. This cost the buyer only the price of the new panels.

Manufacturer support

The builders in the Palm Beach area all agreed that Royall Wall Systems, Inc. does a great job of supporting them when they use the system. We asked every builder we spoke with for the names of other builders who had used the system but didn't like it or quit using it, and they couldn't name any.

The manufacturer takes a lot of work off the builder's hands: assembling and installing the walls, getting fully engineered plans to the building department and dealing with officials there, and checking the engineering and making changes when they're needed. The company also has representatives to answer questions from builders, the trades, and customers.

Beyond these things, builders found they didn't need much extra support. The panels connect easily enough to other construction components that after they're up, there's little need for help.

Other considerations

People agreed that the Royall Walls are quieter than conventional frame. However, the manufacturer has not had their STC measured yet.

They also agreed that construction sites using the system are much neater than others. There's virtually no waste of any kind from the exterior walls. What's trucked in is almost exactly what's needed.

Further information

Royall Wall Systems, Inc.
300-A Royal Commerce Road
Royal Palm Beach, Florida 33411
(407)-689-5398

Superior Wall

Designed originally for basements, the Superior Wall has started appearing above grade as well. It can provide a simple but high-quality structure at only a little more cost than standard frame. And it can do it in days or weeks of less time. Homes built of the panels have sold to cost segment customers who wanted something more solid for their money, and to value buyers satisfied with a simple design.

One builder said to us, "If you took an ordinary stud wall and magically turned the wood parts into concrete, you'd get something pretty close to the Superior Wall." Figure 9-9 contains a diagram of a standard Superior Wall panel. The outside (where the plywood sheathing is on a stud wall) is a continuous face of fiberglass-reinforced concrete about 2 inches thick. Capping the panel at top and bottom are concrete plates, the top serving as a sill plate and the other as a footing or bottom plate. The sill plate has preformed holes for bolting a 2-x sill or another wall above. Just inside the outside face is a 1-inch layer of Dow Styrafoam extruded EPS foam to serve as insulation. Inside of the foam is a steel-reinforced concrete "stud" every two feet o.c.

Each stud is about 2¼ inches wide, 7¼ inches deep, and has a pressure-treated nailer attached to the inside face. The studs are attached to the top and bottom plates, and they're connected to the outside face with metal fasteners that pierce the foam. The studs have preformed holes about every two feet for utility lines and attaching other things to the panel.

The panels are available in standard heights of 4 feet (for use as a frost wall); 8 feet, 2 inches; and 10 feet, with openings as specified by the builder. Some builders arranged for special production of walls with nonstandard dimensions and features when they needed them.

The inventor is Superior Walls of America, Ltd., of Ephrata, Pennsylvania. The company licenses other concrete casting companies to produce the panels to order. The vast majority of the walls to date have been used for basements. They're set on a bed of crushed stone, and then a slab is poured inside. No separate footer is necessary if the soil is sound. However, a growing number of builders are using the panels for above-grade walls as well. Some have built

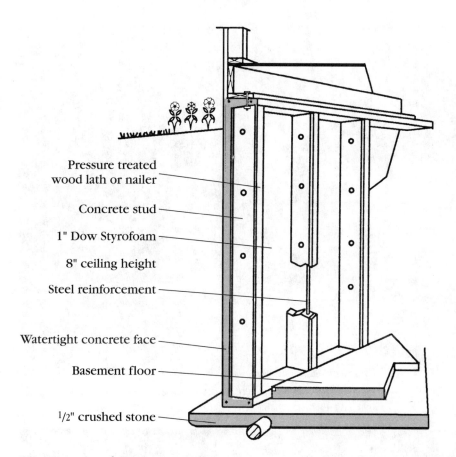

Pressure treated
wood lath or nailer

Concrete stud

1" Dow Styrofoam

8" ceiling height

Steel reinforcement

Watertight concrete face

Basement floor

¹/₂" crushed stone

9-9 *Diagram of a Superior Wall panel.* Superior Walls of America, Ltd.

ranch houses by setting the panels directly on crushed stone or on a conventional foundation, or they've stacked them to form a basement plus first story.

The local Superior Wall manufacturer sends out its own crew (called a set crew) to put up the panels for the builder. When the foundation or bed of stone is ready, the manufacturer's truck delivers the panels, and a crane lifts them into place while the set crew positions them precisely. The photograph in Fig. 9-10 shows this. The crew then bolts the panels together with a specially designed bolting system and installs sealant along the joints. If the Superior Wall is being used for one story only, the set crew leaves at this point, and a concrete crew comes in to pour the slab. The slab is poured against the footer plate, which locks the walls in place and prevents them from pushing in. Figure 9-11 shows the result.

If the panels are stacked, the set crew stops at the top of the first level to wait for the slab and, sometimes, the first floor deck. These strengthen the structure before the second level of panels. The set crew then returns to put the sec-

9-10 *Set crew setting a panel on a crushed stone footing.* Superior Walls of America, Ltd.

9-11 *Completely set Superior Wall panels locked in place with the floor slab.*

ond level of panels on top of the first, bolting them through holes in the plates. Once the panels are set, bolted, and sealed, construction can proceed almost exactly as it does with a standard frame wall.

Superior Wall panels are manufactured in Pennsylvania and New York state and are shipped to neighboring states as well. Superior Walls of America expects to sign up plants in other areas in coming years.

Labor and materials' costs

Table 9-2 gives costs for the system in the middle of the range of figures that builders gave us.

The cost of the panels goes up or down with the number of corners and window openings. The $4.75 number in Table 9-2 applies to a rectangular floor plan with a couple of doors and an average number of windows. According to the manufacturer, a lot of corners can increase the price by almost a dollar a square foot. Each window opening adds about $30, and each door opening about $40. Larger openings (as for sliding doors or garage doors) can add even more. Modifying the panels to stack or other unusual features also add cost although exactly how much varies with the details. You give the specs to the local plant to get an estimate.

Table 9-2. Representative costs for Exterior Walls Constructed with the Superior Wall System[1]

Item		Cost per Square Foot
Superior Wall bill (includes panels, setting, and sealing)		$4.75
Extra insulation		0–.25
None	$ 0	
3½" fiberglass	.15	
5½" fiberglass	.25	
Subtotal, structure and insulation		$4.75–5.00
Exterior siding and finish, materials and labor		$ .50–1.00
Paint only	$ .50	
PC stucco and paint	1.00	
Interior sheetrock and paint	1.50	
Total, including finishes		$6.75–7.50

[1]The costs listed here are in the middle of the range of quotes we received from builders reporting their actual project costs.

Learning costs

All the builders agreed that there's little new for the trades to learn with the Superior Wall. The manufacturer's set crew does almost all the unconventional work when it sets the walls. Except for a few minor details, the trades work on the structure the same way they would work on frame. In no cases did we hear of any subcontractors who charged more than their usual rates to work on a house made of Superior Wall.

The builder himself did have some learning to do. This included contacting the local plant, learning what things have to be specified, and getting to know how the plant's production scheduling works. The builders said that it's "time well spent" to meet with the plant three to four times before your first Superior Wall project. It's also worth meeting with them on later projects where you're trying something you haven't done with the panels before, such as stacking them or making irregular angles. There's more than one way to accomplish most things, so you and the plant need to agree on the details so that everything gets installed in the plant just the way you'll need it in the field.

Buyer reaction

One builder told us, "We were pitching these homes at the lower-income first-time market. The price and location were the most important things that sold them, but the solidness of the construction differentiated them from all the other builders going after that group. And we could do them for just about the same price as frame. They sold fast."

This summed up the stories we heard. Buyers attracted to the houses constructed of Superior Walls above grade tended to have limited budgets, but appreciated quality construction.

Exterior and interior finishes

The panels come with a broom finish on the exterior that's a little coarser than the finish applied to concrete sidewalks. Some builders simply painted this with a good latex made for concrete. Stucco is an inexpensive alternative. It goes on as cheaply as onto any concrete surface, and the rough broom finish makes for good adhesion.

In case you want to attach a nailed or screwed siding, the plants will install furring strips directly into the panels on the exterior surface, flush with the concrete. This costs about $.25 per square foot extra. We didn't talk with any builders who used these, however. Inside, all our builders finished with sheetrock, attaching it to the nailers on the studs.

Design flexibility

The system has been used in structures two stories tall, including the basement. The manufacturer thinks it could go to three. Any wall angle is possible, although the plant will probably charge more for the odd ones. Figure 9-12 shows some irregular angles made with the system. A true curve isn't available, but some buildings have used a series of short panels that comes close. The large number of angles adds to the cost, however, as noted previously.

9-12 *Highly irregular walls constructed out of Superior Wall panels.*
Superior Walls of America, Ltd.

None of our builders asked for curved openings of any type. A representative of Superior Walls of America said that they're unaware of one ever being made, but "We and the plants—we'll try a lot of things if you want to pay for it, but for first-time stuff the cost can be pretty high."

R-value and energy efficiency

The standard 1 inch of extruded EPS that's installed in the panels is rated at R-5. This insulation is broken by the sill and bottom plates, and by the connectors that run between the face of the panel and the studs. Dow has calculated the parallel-path R-value of a full panel at 7.5.

The studs provide room for up to 7¼ inches more insulation. Everyone we talked with who added insulation here put in fiberglass. By filling this cavity with fiberglass, you can add over 20 to the insulation R-value of the wall. This extra insulation is broken by the studs and top and bottom plates. The manufacturer has not done separate calculations of the R-value of a full panel with insulation between the studs.

The concrete of the 8-foot panels has a weight of about 36 pounds per square foot. From this we estimate the thermal mass of the walls at about 8 Btu/sf/degree, more than enough to get R-value credit from most energy codes.

Thermal mass

The concrete of the studs of the Superior Wall are inside the standard insulation. The thermal mass of the studs works out to about 1.5 Btu/square foot of wall area/degree. We didn't find any houses built of Superior Walls that used solar heating or load leveling.

Water resistance

The concrete of the Superior Wall panels is designed to be water resistant. The panels are made of concrete with a compressive strength of 5000 pounds per square inch. The American Concrete Institute states that concrete of 4000 psi or more can be considered "watertight." None of our builders had experienced any leakage through their panels, whether they were below or above grade. According to the manufacturer, out of the roughly 15,000 basements made with the system, they get three to five complaints a year of leakage through a joint between panels. They say it happens when the set crew didn't seal the joint adequately. When it happened, the plant sent out a crew to repair the leak.

The manufacturer claims to have had no reports of leakage through walls above grade. Because the builders we interviewed covered their above-grade walls with paint or stucco and paint, they had this additional water barrier.

Maintenance required

None of our builders had performed any maintenance on their above-grade Superior Walls yet. We assume that they'll have to be repainted about as often as other concrete walls, every six years or so, compared with every four for wood sidings.

The only other form of maintenance we've heard about was repair of leaking joints below grade, mentioned previously. In these cases the plants repaired the leak at no charge by sending out a crew to dig around the joint and reseal. According to a staff member of Superior Walls of America, "Once they do this a couple of times, they figure out it's a lot cheaper to do it right the first time." You might want to stress the importance of a good seal to the plant just to be safe. You might also want to check on the plant's warranty. Most of them warranty the walls and their installation for two years against defects in materials and workmanship. If leaks develop during that period because of the sealant work, they have to fix them.

Disaster resistance

The walls have not yet been put through a fire test. However, an engineer working for the manufacturer has estimated that the unfinished panel would qualify as a one-hour fire wall, and if the stud side were covered with two layers of sheetrock this would increase to two hours.

Superior Walls of America has recently received approval for above-grade use of the system in Florida from the State Department of Community Affairs, which requires houses to withstand 160 mph winds. The company's engineer calculates that a typical ranch house made with 10-foot walls with 2 feet below grade and a slab poured between the walls to tie them together could withstand winds of an incredible 270 mph. According to the manufacturer, when a tornado touched down in Limerick, Pennsylvania and demolished the above-ground structures of several houses, some of the traditional foundations were damaged, but none of the Superior Wall foundations were.

The panels have also not been used in high seismic areas, and they haven't been tested for earthquake resistance. The manufacturer says that after quakes in the Reading, Pennsylvania area, they found that some other foundations had been damaged, but none of the foundations of the Superior Wall showed any.

Product availability

The Superior Wall is manufactured in six plants in Pennsylvania and two more in New York state. These plants have shipped as far as the Plains States, but that's rare because of transportation costs. The shipping charges work out to about $.12 per square foot of panel for each 25 miles traveled. None of our builders had panels shipped more than 100 miles.

Because the plants all have backlogs, you have to order panels in advance. Some plants claim to need as long as eight weeks to deliver, although they say they're trying to get that down to four. But remember that you can place an order without knowing all the details. Some builders just put their name on the production list as soon as they knew when they would need panels and about how many. The plant didn't actually need blueprints until roughly 10 days before the requested production date.

Labor availability

The builders that used the system said they didn't need any unusual labor, and their existing crews didn't need much training. The plant's set crew does virtually all of the unusual tasks. The manufacturer supplies an instruction manual that goes over details such as how to attach sill plates. But these are generally the same as normal construction practice.

Building department approval

The system has a BOCA Research Report (No. 89-53) that covers its use in basements. Nonetheless, in the homes we studied, the building department accepted it for use above-grade as well without extra documentation. As one builder put it, "In our part of the country, the officials have already seen this before in base-ments and they're comfortable with it. Maybe in other parts of the country it would be different."

If you need to have an engineer analyze and stamp your plans, the plant that sells the panels can direct you to one who's already familiar with the sys-tem. They claim that since plans rarely have to be changed significantly, the cost is usually in the $100–300 range.

Required calendar time

Everyone agreed that the walls go up fast. One builder said "That was one of the big advantages of this system to us. It took maybe two weeks off our construc-tion schedule."

The set crew takes about a half day to set one level of walls. If you're stack-ing up more than one story, they have to wait for the deck to be built, so total construction time can be as little as 1½ days (½ day for the first-story panels, an-other half for the deck, and another half for the second floor), or longer if you have waiting time between crews.

Roof and interior work can begin almost immediately after the set crew leaves. However, the manufacturer does recommend that for full basement homes, you wait until the slab is poured and backfilling is done.

The panels can go up in almost any weather. Whether you're using a gravel base or a poured footing, the installation can occur when the ground is frozen. It's possible to erect the panels in the rain, although the set crews usually don't go out if they expect a hard one. When they do install in the rain, they have to come back later to seal beacuse the concrete has to dry for at least a couple of hours before applying sealant.

Crew coordination

There's almost no coordination that the builder has to do between the set crew and the other crews. The other crews come in after the walls are up and work off of them just about the same way they do with a conventional stud wall. The two places where coordination is necessary is in specifying the openings and building intermediate floor decks.

The builders confirmed that it's important to have all the large openings specified correctly in the plans before the walls are made. So you might have to talk with your suppliers and interior carpenters to make sure everyone agrees on the style, sizes, and positions of windows, doors, etc.

When you're stacking walls, the panels used on the second level are usually specially made to take joists on a ledge or in pockets. As with openings, this is something that has to be worked out in advance. You might want to check with your carpenters to make sure the way the panels are made and the way they plan to build the floor match. In addition, the builder has to schedule the set crew and carpentry crew to alternate at the job site.

Making connections

Roofs fasten to a 2-x top plate that's bolted to the concrete top plate of the Superior Wall through preformed holes. Intermediate floor decks are a little trickier. The builders we spoke with met with their plant about how to handle them, and the plant custom-designed a second-level panel to do the job. The panel had either a narrower bottom plate that left a 4-inch shelf or notches that served as pockets. In either case, the joist ends up resting on the top plates of the panels below. This type of arrangement is necessary because you can't set the joists on the bottom plate of the upper panels. Their top surfaces slope inward, so they're not a steady base.

Superior Walls of America suggests two other methods for supporting floor decks. One is simply to build out the front and back walls of the lower level with 2-x-4 frame bearing walls and rest the joists on these. The other is to add extra beam pockets to the side walls so that you can set two extra beams at the top of the first level, one near the front wall and one near the back. The extra beams support the ends of the joists.

The plant installs pressure-treated 2-x bucks flush with the outside surface of concrete around all door and window openings. So you can install the doors and windows in nearly any way you install them on frame.

Interior walls, cabinets, and fixtures nail or screw into the nailers on the studs just as they attach to a frame wall. Where you need to connect to points between the studs, you can set a horizontal block that runs between two studs and toenail it to the nailers.

Utilities installation

Most lines run in the wall, just as they do in stud walls. There's plenty of space in the cavity to go vertically with even the largest vent stack. If you ask for it in advance, you can also get a sleeve installed in the top plates to go up through them.

Small lines go horizontally through the preformed holes in the studs. Going horizontally with larger lines is tougher because the studs are designed to bear a certain load. Putting large holes in them can give you an unsafe structure. Some plants will cast large holes through the studs, but sometimes they recommend going above the top plate before going sideways.

To make holes to the outside, you can drill through the two inches of exterior concrete with a masonry bit or tap a hole with repeated hits from a hammer. Or you can get sleeves cast in, if you tell the plant where you want them in advance.

Change flexibility

One builder said, "We never had to make any field changes in the structure. That was the point of having so many meetings with the plant ahead of time."

Another who had done a large number of homes with Superior said that a few times they'd moved a window. "We just sent out a crew with a saw and a lot of masonry blades. They rip out the buck, recut the hole, and put in a new buck. It's a hassle, but you deal with it, just like with frame." It reportedly took a crew of two about half a day to do the job, at a cost to the builder of about $200.

Manufacturer support

The builders felt they got plenty of support from their plants. The plants were willing to hold as many meetings as necessary to get even unusual orders planned out fully. They provided "plenty of documentation" for the builders and building inspectors. In some cases they also made presentations to the local building department so that the inspectors understood the system.

They also did presentations for the builders. But most of the builders we spoke with were so familiar with the system from using them in basements earlier that they didn't really need more training. They just needed some planning on how to use it above grade.

Other considerations

Everyone agrees that the job site of a Superior Wall project is unusually neat. After the set crew installs the exterior walls, there is almost no debris (no cutoffs, no nails, no glue drippings).

People who have been inside the finished houses say they're noticeably quieter than frame homes. However, the manufacturer has not had the panels tested for their STC yet.

Further information

Superior Walls of America, Ltd.
P.O. Box 427
Ephrata, PA 17522
(717)-626-WALL

On the horizon

Tierra Homes of Oroville, California has recently introduced its own concrete wall panels. They're pictured in Fig. 9-13. In contrast to other brands, Tierra's panels have the structural concrete inside and a layer of foam insulation outside. This gives the finished home a high thermal mass inside the insulation envelope. The houses constructed so far have taken advantage of this with south-facing windows that bring in heat during the winter. That heat gets stored in the concrete. The energy savings and comfort level of these homes have been impressive. The exterior can be finished with most conventional sidings, which cover and protect the foam. Tierra is now marketing the panels to builders in North-

9-13 *Tierra Homes wall panels on a truck ready for delivery.* Tierra Homes

ern California, as well as offering to license manufacturing rights to other companies that want to make and sell panels in their own area.

Dallas Harris Real Estate Construction of North Carolina has recently begun offering builders a tilt-up system called DailRoc. In the patented DailRock system, the crew assembles, on site, a grid of steel rebar and a few steel plates inside a mold for a wall panel. When the steel is in place, concrete goes in the mold and cures to form a reinforced concrete panel. The panels are lifted into place and welded to one another and to the foundation. After the panels are installed, their insides are lined with a vapor barrier, rigid foam insulation, and gypsum wallboard. According to the manufacturer, the result is an extremely strong, durable, energy-efficient structure for about the same cost as frame. Dallas Harris is interested in entering licensing agreements with other builders who would also like to offer the system.

Further information

Tierra Homes
6898 Lincoln Boulevard
Oroville, CA 95966
(800)-373-9930

Dallas Harris Real Estate Construction
P.O. Box 531
7208 Wrightsville Ave.
Wrightsville Beach, NC 28480
(910) 256-4475

10

Getting started

Aside from the logistics of supervising workers and putting the structure together, our builders had some tips for making the construction of your first few CHS homes a successful experience. Several of these recommendations came up often enough that we thought builders evaluating the new systems would be well advised to consider them.

Pace yourself

Like anything new, building with a CHS for the first time is full of unexpected surprises. Just after he did his first house with stay-in-place forms, one builder told us, "I talked with four different guys who did this before. They all said they spent a couple thousand extra dollars the first time because they made mistakes, everybody worked slower, and things like that. But I figured I could save that money because I was talking to them and I'd figure out all the mistakes ahead and avoid them. I mean, I watched them put up a couple of houses, I called them constantly during construction with questions and problems—everything I could have done. But still a hundred little things came up that I'd never thought of. And darned if after it all I didn't wind up spending a good 2000 dollars extra on that job."

The message the experienced CHS builders gave us was not to get frustrated. The first house is a learning experience, and you should consider it one. Go ahead and think hard about everything you do, ask questions, try to anticipate problems. But don't expect to avoid the learning costs that everybody else had to go through. The real measure of how well you did is the cost of the second house. That tells how well you learned.

In case it isn't already obvious, a new CHS makes the most sense for someone interested in building more than one house with it. A builder who does only one gets the frustrating part (the mistakes of the first house) without the more enjoyable and profitable part (the smooth operation on later houses). We met a couple of builders who planned to build only one house with a new system, did it, and were satisfied with the experience. But they were unusual. They simply wanted something unique and flashy to get attention in a parade of homes or high-profile development. And they subcontracted almost all of the work to another builder who had experience with the system involved.

The builders who came through their first house most satisfied were the ones who had a clear idea of what type of house they wanted to sell, who would buy it from them, and why those people would want it. In short, they had a vision for the business they wanted to create, and they understood how the CHS they had chosen would help them create it. To these people, the problems they encountered on the first house were just minor bumps in a long road. The problems were even helpful in one way: they taught the builder more about how the system worked and why. This made it possible to do the job all the better on later projects.

The happiest builders were also those who undertook their first new CHS house when they had some slack, when they had a little extra money, and when there weren't old customers lined up asking for a lot of work to get done. If the builder had slack, when problems arose he didn't feel pressured. Instead he could sit back and think through the best way to handle the situation. As one put it, "This is a thing to do when you have a little time and money to spare."

Try a basement first

Several builders eased into building with concrete by using a new CHS on a small section of the house the first time. They could learn a lot about the system by using it, yet time or cost overruns would be limited.

Basements are a good part of the house to start with. They have few corners, windows, or doors, so you're not overwhelmed with a lot of complexity when you're using an unfamiliar system for the first time. Because basements are mostly backfilled, you also don't have to do much exterior finish work on them. And you can usually get some more money for them to help cover the extra costs of the premium wall. Nearly all the CHSs include insulation and some means of finishing the interior inexpensively. So for only a little cost over and above the basic cost of the system, several of our builders used one of the CHSs to make a finished basement. And they succeeded in charging extra for it.

But any section of a house can do the trick: a wing, an addition, or even a frost wall or retaining wall. They'll all give you some experience with the system at low risk.

Call the newspaper

Once they were confident they could make an impressive house, a lot of our builders found it both easy and worthwhile to get publicity. The usual method was to call the local paper when the structure above grade was just getting started. To get attention from reporters, they used exciting expressions to describe what they were doing—things like, "stacked up like Lego blocks," "held together with Super Glue," "made out of recycled foam coffee cups." They waited until the structure was underway so that they'd be able to show reporters the guts of the system and some construction in action.

The builders who tried this were almost always successful in getting at least one article in a medium-sized town paper. They reported getting up to 100 calls from people who had read the article, with most coming in the first three weeks after it appeared. Typically those 100 calls led to orders for 2–4 actual new houses. The calls and the follow-on projects that came out of them also generated word-of-mouth that led to more inquiries later.

Be proud

The builders we interviewed decided to try a new concrete homebuilding system for a combination of three reasons: they wanted an edge over their competition to increase sales and profits, they found new technology interesting, and they believed the system they'd chosen was superior to conventional construction.

A couple of builders felt sheepish about talking about superior construction. They thought they would sound "corny" if they told buyers their houses were better than the ones other contractors built. But most builders said that this was a mistake. "The quality of the construction is maybe the biggest thing you have going for you," one said. "If you don't tell people about it, nobody will." Another said, "It's really motivating to think that you're bringing a better product to the market. You feel like you're going to make your money by being the quality firm. Like you're leading the way and setting the standard for houses of the future."

Good luck.

Directory of product manufacturers

Exterior insulation and finish product manufacturers

ACROCRETE INC.
3009 NW 75th Avenue
Miami, FL 33122
Phone (800) 432-5097
FAX (305) 591-1497

COREV AMERICA INC.
11620 Brittmore Park Drive
Houston, TX 77041
Phone (713) 937-3437
FAX (713) 937-9765

DRYVIT SYSTEMS, INC.
One Energy Way
West Warwick, RI 02893
Phone (401) 822-4100
FAX (401) 822-4510

FINESTONE
Division of Simplex Products
Adrian, MI 49221
Phone (517) 263-8881
FAX (517) 263-2835

KEYSTONE SYSTEMS, INC.
22 Denlar Drive
Chester, CT 06412
Phone (203) 526-9515
FAX (203) 526-5831

PAREX, INC.
P.O. Box 189
Redan, GA 30074
Phone (404) 482-7872
FAX (404) 482-6878

PLEKO PRODUCTS
2646 Alliston Court
Columbus, OH 43220
Phone (614) 451-7282
FAX (614) 451-7392

SENERGY INC.
1367 Elmwood Avenue
Cranston, RI 02910
Phone (401) 467-2600
FAX (401) 941-7480

STO CORPORATION
6175 Riverside Drive, S.E.
Atlanta, GA 30331
Phone (404) 346-3666
FAX (404) 346-3119

THERMOMASS
(Plastic pins used to connect
stucco mesh over foam)
525 E. 2nd Street
Ames, IA 50010
Phone (515) 232-1748
FAX (515) 232-0800

THORO SYSTEM PRODUCTS
7800 N.W. 38th Street
Miami, FL 33166
Phone (305) 592-8100
FAX (305) 592-9760

Mortared block system manufacturers

HEBEL SOUTHEAST
Brand name: Hebel Wall System
4000 Cumberland Parkway
Suite 100B
Atlanta, GA 30339
Phone (404) 344-2897
FAX (404) 691-2843

SPARFIL BLOC FLORIDA
Brand name: Sun Block
PO Box 270336
Tampa, FL 33688
Phone (813) 963-3794
FAX (813) 963-3794

SUPERLITE BLOCK
Brand name: Integra Wall System
4100 W. Turney
Phoenix, AZ 85019-3327
Phone (602) 352-3500
FAX (602) 352-3813

Mortarless block system manufacturers

INSULATED MASONRY
SYSTEMS, INC.
Brand name: IMSI Block
7234 East Shoeman Ln., Ste 1
Scottsdale, AZ 85251
Phone (602) 970-0711
FAX (602) 970-1243

INTRALOCK CORPORATION
Brand name: Intralock Block
1001 NW 12th Terrace
Pompano Beach, FL 33069
Phone (305) 942-0000
FAX (305) 942-7700

SPARFIL INTERNATIONAL, INC.
Brand name: Sparfil Wall System II
376 Watline Avenue
Mississauga, Ontario L4Z 1X2
Phone (905 507-1163
FAX (905) 8990-7784

Poured-in-place system manufacturers

AAB BUILDING SYSTEMS
840 Division St.
Cobourg, Ontario Canada K9A 4G9
Phone (905) 373-0004
FAX (905) 373-0002

AMERICAN CONFORM INDUSTRIES
Brand name: Smart Block
1820 South Santa Fe St.
Santa Ana, CA 92705
Phone (800)-CON-FORM
FAX (714) 662-0405

AMERICAN POLYSTEEL FORMS
5150 F Edith NE
Albuquerque, NM 87107
Phone (800) 977-3676
FAX (505) 345-8154

ENERG CORP.
4203 West Adams
Phoenix, AZ 85009
Phone (602) 470-0223
FAX (602) 438-7678

ENERGRID
6421 Box Springs Blvd.
Riverside, CA 92507
Phone (909) 653-3346
FAX (909) 653-0986

FEATHERLITE BUILDING PRODUCTS
301 West Hills Rd.
New Canaan, CT 06840
Phone (203) 966-2252
FAX (203) 966-0246

GREENBLOCK WORLDWIDE CORP.
P.O. Box 749
Woodland Park, CO 80866
Phone (719) 687-0645
FAX (719) 687-7820

I.C.E. BLOCK
570 South Dayton-Lake
View Rd.
New Carlisle, OH 45344
Phone (800) 423-2557
FAX (513) 845-9837

KEEVA INTERNATIONAL, INC.
1854 North Acacia St.
Mesa, AZ 85213
Phone (602) 827-9894
FAX (602) 827-9697

LITE-FORM, INC.
PO Box 774
Sioux City, IA 51102
Phone (800) 551-3313
FAX (712) 252-3259

PERMAFORM U.S.A., INC.
7 Doubleday Lane
Ridgefield, CT 06877
Phone (203) 438-8357
FAX (203) 431-0177

DISTRIBUTION OF POLYCRETE, INC.
7390 Henri Bourassa Blvd. East
Anjou QC H1E 1P2
Phone (514) 493-2854
FAX (514) 493-1904

R-FORMS, INC.
10999 Prosperity Farms Rd.
Palm Beach Gardens, FL 33410
Phone (407) 624-2515
FAX (407) 624-9838

REDDI-FORM, INC.
593 Ramapo Valley Rd.
Oakland, NJ 07436
Phone (800) 334-4303
FAX (609) 405-1987

Shotcrete system manufacturers

COROTHERM BUILDING SYSTEMS
P.O. Box 978
Guntersville, AL 35976-0978
Phone (205) 582-0808
FAX (205) 582-0728

IMPAC INTERNATIONAL
Brand name: Thermal-Impac Panel
16441 Orange Way
Fontana, CA 92335
Phone (800) 227-9591
Fax (909) 829-3089

INSTEEL CONSTRUCTION SYSTEMS, INC.
Brand name: 3-D Panels
2610 Sidney Lanier Drive
Brunswick, GA 31525-9003
Phone (800)-545-3181
FAX (912) 264-3774

TRUSS PANEL SYSTEMS
Brand name: Truss Panels
Estate Pastory #7, St. John
U.S. Virgin Islands 00830
Phone (809) 776-6237
FAX (809) 779-4811

Panelized system manufacturers

ROYALL WALL SYSTEMS INC.
Brand name: Royall Wall
300-A Royal Commerce Road
Royal Palm Beach, FL 33411
Phone (407) 689-5398
FAX (407) 689-0407

SUPERIOR WALLS OF AMERICA, LTD.
Brand name: Superior Walls
P.O. Box 427
Ephrata, PA 17522
Phone (717) 626-9255
FAX (717) 626-7319

TIERRA HOMES
Brand name: SolarCast
6898 Lincoln Blvd.
Oroville, CA 95966
Phone (916) 534-6572
FAX (916) 534-6073

Metal stud insulation system manufacturers

GRACE CONSTRUCTION PRODUCTS
Brand name: Thermo-Stud Wall Insulation System
62 Whittemore Ave.
Cambridge, MA 02140
Phone (617) 876-1400
FAX (617) 498-4419

Concrete block
pre-insulation manufacturers

BLOCFIL CO.
Brand name: Blocfil (standard)
Division of Patek Investment Corp.
PO Box 405
Chillicothe, MO 64601
Phone (800) 441-1050
FAX (816) 646-3610

FABRICATED PACKAGING MATERIALS, INC.
Brand name: Fabri-Core (standard)
296 Quarry Rd.
PO Box 306
Lancaster, OH 43130
Phone (614) 687-5934
FAX (614) 687-3671

GRACE MASONRY PRODUCTS
Brand name: Korfil (standard)
Hi-R (specialty)
62 Whittemore Ave.
Cambridge, MA 02140
Phone (617) 876-1400
FAX (617) 498-4419

INSUL BLOCK
Brand name: PolyCore (standard)
PolyLok (specialty)
P.O. Box 1162
West Springfield, MA 01090-1162
Phone (800) 235-0061
FAX (413) 781-0072

THERMALOCK PRODUCTS, INC.
Brand name: ThermaLock
(specialty)
162 Sweeney Street
North Tonawanda, NY 14120-5908
Phone (800) 537-6363
FAX (716) 695-6000

WEST MATERIALS, INC.
Brand name: Enerblock (specialty)
101 West Burnsville Parkway
Burnsville, MN 55337
Phone (612) 890-3152
FAX (612) 890-9341

Foam-in-place insulation manufacturers

CP CHEMICAL, INC.
25 Home St.
White Plains, NY 10606
Phone (914) 428-3636
FAX (914) 428-2517

.JESCO
38-10 Urby St.
Florence, SC 29502
Phone (803) 665-5350
FAX (803) 665-5351

POLYMASTER, INC.
10431 Lexington Drive
Knoxville, TN 37932
Phone (615) 966-3005
FAX (615) 675-3300

THERMAL CORP.
Rt. 3 Highway 34 West
Mt. Pleasant, IA 52641
Phone (319) 385-1535
FAX (319) 385-1540

THERMO TEC
1263 Sawdust Trail
Kissimmee, FL 34744
Phone (407) 847-7463
FAX (407) 847-3689

Pour-in-place insulation manufacturers

GRACE CONSTRUCTION PRODUCTS
62 Whittemore Ave.
Cambridge, MA 02140
Phone (617) 876-1400
FAX (617) 498-4419

WESTERN FORMS, INC.
6200 Equitable Road
Kansas City, MO 64120
Phone (800) 821-3870
FAX (816) 241-6877

Glossary

AAC Abbreviation for aerated autoclaved concrete.

ACC Abbreviation for autoclaved cellular concrete.

aerated autoclaved concrete A specially manufactured concrete that contains a large number of fine air bubbles.

aggregate Material in larger pieces mixed with Portland Cement and water to form concrete. Includes sand, stone, and sometimes other materials.

architectural block Concrete block that is specially manufactured to give one or both face shells a particular appearance in its color, shape, or texture.

autoclaved cellular concrete See aerated autoclaved concrete.

Basic Building Code The model code most widely used by states and localities in the Northeast and Midwest.

bead board Sheet stock made of molded EPS.

blowout The bursting of concrete out of the concrete forms into which it was poured.

BOCA Code Nickname for the Basic Building Code.

BOCA Abbreviation for Building Officials and Code Administrators International.

bond beam block See U-block.

bond beam A horizontal band of concrete and rebar inside the walls of a structure. Extends completely around the exterior walls to tie the structure together.

British thermal unit The amount of heat necessary to raise 1 pound of water 1 degree Fahrenheit.

Btu See British thermal unit.

buck A wooden frame set into an opening in a concrete wall so that a window or door can be fastened to the buck, and therefore to the wall.

Building Officials and Code Administrators International An organization of primarily building officials that writes and modifies the Basic Building Code.

cavity-insulated block A form of wall construction in which insulation is placed inside the cavities of a block wall.

cavity An open space inside a concrete block.

compressive strength The ability of a material to withstand compression (or "a load"), usually measured in pounds per square inch (psi).

concrete pump A pump designed to push concrete through a hose for pouring into form work or shooting onto a frame.

conduit Flexible metal or plastic tubing inserted into a wall to serve as protection for electrical cable placed inside the tubing.

corner block A block designed to be stacked at the corner of a wall.

cost segment The segment of buyers who want a home at especially low cost and are willing to accept products and building practices below standard quality to get it.

course A horizontal row of blocks at the same level in a wall.

cut nail A type of heavy nail that can be hand-driven into concrete.

cut-down web A web of a block that does not extend to the top of the block, but instead ends part way up.

density The weight per unit volume of a material, which in plastic foams is higher when the air bubbles in the foam are smaller and more numerous.

double-wythe wall A wall consisting of two separate walls of masonry with an air gap in between.

drilled fastener A fastener attached into a hole that has been drilled for that purpose.

driven fastener A fastener attached with impact, as from a hammer.

dry cup method A method of determining the water vapor permeance of a material by placing it over an empty container in a humid room and measuring how fast water vapor passes through the material and into the container.

dry-mix shotcrete See dry shotcrete.

dry shotcrete Shotcrete that is forced through two separate lines, one containing the dry ingredients and one containing the water, and mixed at the nozzle as it is shot from the hose.

8-inch module The method of designing buildings so that breaks in the walls (corners, openings, etc.) occur at some multiple of 8-inch dimensions.

8-x-8 reinforcing grid The pattern of reinforcing formed by placing one vertical steel reinforcing bar every 8 feet in a wall and one horizontal steel reinforcing bar every 8 feet.

elastomeric grout A grout with a high polymer content that gives it the ability to compress, flex, or stretch without tearing or cracking.

elastomeric paint A paint with a high polymer content that gives it the ability to flex or stretch without tearing or cracking.

energy segment The segment of buyers who want a home that is more energy efficient than standard construction and are willing to pay more for it.

EPS See expanded polystyrene.

evaluation report A report written by a model code organization that verifies that the tests performed on a particular building product or system, when used according to specifications, meets the requirements of the organization's model code.

expanded polystyrene A plastic foam made of polystyrene filled with tiny gas bubbles.

exposed block See architectural block.

exterior-insulated block A form of wall construction in which insulation is placed on the exterior face of a block wall.

extruded EPS A seamless EPS foam made by extruding EPS continuously through a die.

face shell A face (either inside or outside) of a block.

fire resistance rating The length of time that a wall has been subjected to the fire of a fire wall test without failing the first two parts of the test (exposure to fire without structural failure and without allowing extreme heat to pass through the wall).

fire wall rating The length of time that a wall has been subjected to the fire of a fire wall test without failing any of the three parts of the test.

fire wall test A laboratory test of the ability of a wall to: withstand fire without failing structurally and prevent extreme heat from the fire from passing through the wall.

form oil An oil applied to the inside surfaces of concrete forms so that they can be more easily removed after the concrete hardens.

4-×-4 reinforcing grid The pattern of reinforcing formed by placing one vertical steel reinforcing bar every 4 feet in a wall and one horizontal steel reinforcing bar every 4 feet.

four-high See half-high.

glazed block Concrete block that has a glazing on one side to provide a smooth, colored surface.

grid system A type of stay-in-place form system consisting of molded foam blocks stacked up to act as the concrete form.

gross wall area The area of a wall, including the area of any openings (such as doors and windows) in the wall.

ground-face block Concrete block with a face shell that has been ground smooth to give an appearance similar to polished stone.

grout A thin form of concrete with only very small aggregate, used mostly to fill small cavities for reinforcing purposes.

guarded-hot-box R-value The R-value estimated for a wall by the guarded hot box test.

guarded-hot-box test A test to estimate the R-value of a wall or material by keeping one side of the wall or material cooler than the other and measuring how fast heat passes through.

gunite See dry shotcrete.

half block A concrete block that is 8 inches long, instead of 16 inches.

half-high block Concrete block that is 4 inches high, instead of 8.

hard-coat stucco See polymer-modified stucco.

hurricane straps Metal straps used to connect roofing members to the wall beneath them.

ICBO Code Nickname for the Uniform Building Code.

ICBO Abbreviation for International Conference of Building Codes.

insert A cut or molded piece of foam placed into the cavities of concrete block to act as insulation.

insulation R-value The R-value estimated for a wall by taking the R-value of the insulation in the wall at some representative point.

interior-insulated block A form of wall construction in which insulation is placed on the interior face of a block wall.

International Conference of Building Officials The model code most widely used by states and localities in the western United States.

isothermal-planes R-value The R-value estimated for a wall by an engineering formula that takes into account the sideways movement of heat within the wall.

J-bolt A bolt with threads at one end and a crook in the shank at the other, designed to be anchored into concrete for connection of some other structural member to the concrete.

L-angle A metal plate bent approximately in the center to connect two structural members at right angles.

L-channel Metal channel with a cross-section shaped like the letter L.

latex-based emulsion A coating material made principally of latex, used for dampproofing walls.

learning costs The extra costs that one incurs in learning to do some unfamiliar task efficiently.

ledger A horizontal line of structural lumber fastened to a wall to serve as a connection point for other structural members, usually floor joists.

left-hand corner block A corner block designed to extend from the corner into the wall that lies to the left of the corner, as viewed from inside the corner.

lift In a concrete pour, one pass around the walls of the structure to fill the form work up part way.

lintel block A concrete U-block, or a stay-in-place grid block, with cut-down webs to allow grout or concrete poured into the wall to flow sideways.

market segment A group of home buyers with similar preferences.

masonry nail A nail designed to be driven into concrete.

masonry screw A screw designed to be driven into a predrilled hole in a masonry or concrete wall.

masonry window A window designed to fit into and fasten to a masonry wall.

mass-corrected R-value The R-value estimated for a wall by adjusting some other estimate of its R-value to take account of the thermal mass effect.

model code A building code written as a starting point for states and localities in writing their own codes.

molded EPS An EPS foam shaped by filling a mold with beads of EPS and heating to fuse them together.

monolithic pour A pour of concrete done all at one time, avoiding any seams between wet and hardened concrete.

mortared block Concrete blocks assembled into a wall by placing mortar between them to bind them and keep them level.

mortarless block Concrete blocks assembled into a wall by stacking them directly on one another without mortar and (usually) binding them after stacking with a surface bonding material on the outside surfaces or grout poured into the cavities.

nominal wall R-value The R-value estimated for a wall by adding the R-values of each layer of material (insulation, sheetrock, siding, and so on) of the wall.

nonproprietary Not owned or patented by any one company.

o.c. Abbreviation for on-center, which designates that measurements between objects (such as studs or rebar) are being taken from the center of one to the center of the next.

panel system A type of stay-in-place form system using flat foam sheets held usually a constant distance apart with plastic ties to act as the forms.

panelized Consisting of large panels that are hoisted into place to serve as the walls of a building.

parallel-path R-value The R-value estimated for a wall by averaging the R-values for every straight path through the wall.

PB stucco See polymer-based stucco.

PC stucco See portland cement stucco.

perm A measure of the water vapor permeance of a material equal to 1 gram of water vapor passing through each square foot of the material per hour.

PM stucco See polymer-modified stucco.

pocket An opening left in the side of a wall to hold some structural member, usually the end of a joist or beam.

polymer-based stucco Stucco containing large amounts (about 50 percent) of acrylic for flexibility.

polymer-modified stucco Stucco containing small amounts of acrylic for flexibility.

portland cement stucco Stucco containing no acrylic.

post-tensioning The tightening of tension rods in a wall after it has been erected to serve as reinforcing.

pour The pouring of concrete or grout into forms or a wall of concrete block.

poured-in-place concrete Concrete poured into form work placed in the position of the final walls to form the structure of a building.

poured-in-place foam Foam injected into the cavities of a concrete block wall to act as insulation and (sometimes) an air and water barrier.

powder-actuated pin A fastener driven into concrete, consisting of a pin shot from a gunpowder cartridge (like a bullet) with a special gun.

precast construction Construction using precast concrete panels for the walls.

precast Shaped in forms or molds in a plant before shipment to the job site.

proprietary Owned or patented by a single company.

pump mix A special mixture of concrete designed to work well in a concrete pump.

quality segment The segment of buyers who want a home with premium products or building practices and are willing to pay more for them.

R-value A measure of the thermal resistance of a material or wall, set equal to one when 1 square foot of the material or wall allows 1 Btu to pass each hour when there is a 1 degree temperature differential across the material/wall; set equal to 2 when ½ Btu passes each hour, and so on.

radius wall A curved wall.

rebar See steel reinforcing bar.

reinforcing bar See steel reinforcing bar.

reinforcing grid The pattern of vertical and horizontal steel reinforcing bars placed in a wall.

release agent A substance applied to the inside surfaces of concrete forms so that they can be more easily removed after the concrete hardens.

research report See evaluation report.

ribbed block Concrete block with a face shell molded into a series of vertical ribs.

Richter scale A scale for measuring the severity of an earthquake.

right-hand corner block A corner block designed to extend from the corner into the wall that lies to the right of the corner, as viewed from inside the corner.

running bond The pattern of stacking blocks in a wall so that each block is offset sideways from the one below it by one-half block.

s-i-p Abbreviation for stay-in-place.

safety segment The segment of buyers who want a home capable of resisting disasters (such as fire, high winds, and earthquake) and are willing to pay more for it.

SBCCI Abbreviation for Southern Building Code Congress International.

scored block Concrete block with one or more vertical scores in its face shell to give the appearance of multiple, smaller blocks.

segment See market segment.

seismic zone An area of the United States that has been rated for likelihood and severity of earthquakes on a scale from 0 to 4.

seismic To do with earthquakes.

set crew The crew that sets panelized walls into their final place at the job site.

set The setting of panelized walls into their final place at the job site.

sf Abbreviation for square foot.

shake-table test A test to estimate the earthquake resistance of a wall or structure by fixing it to a large platform that is shaken hydraulically. The wall is then checked for structural failures.

shelf The ledge formed in a wall where its thickness is reduced, usually occurring at the top of a floor, such as the basement.

shotcrete Fine-aggregate concrete sprayed onto a steel mesh or other substrate through a hose under pressure from a concrete pump.

sill block A special concrete block, smaller than standard block and without cavities, designed to sit atop the bottom of a window opening to form a sill.

slump block Concrete block made with a thin concrete so that it sags, to give it the appearance of being handmade.

slump The thinness of a batch of concrete, measured as the number of inches a 1-foot cone of the concrete sags after the cone-shaped mold is removed.

soft-coat stucco See polymer-based stucco.

sound transmission class A measure of the ability of a material or wall to stop the passage of sound equal to the decibel difference between a sound's volume on the side of the wall on which it originates and the opposite side.

Southern Building Code Congress International An organization of primarily building officials that writes and modifies the Standard Building Code.

Southern Building Code Nickname for the Standard Building Code.

split-face block Concrete block with a face shell that has been split off to form a rough, stone-like surface.

stack bond The pattern of stacking blocks in a wall so that each block is directly above a block in the course below.

Standard Building Code The model code most widely used by states and localities in the southeastern and south-central United States.

stay-in-place forms Forms for poured-in-place concrete that are designed to be left in place after pouring, usually to serve as insulation and sometimes also to provide an air and water barrier, a backing for siding, and a fastening surface for siding and other attached components.

STC Abbreviation for sound transmission class.

steel reinforcing bar Steel bar designed to be placed inside blocks or concrete forms and surrounded with concrete or grout for the purpose of reinforcing the wall against side and uplift forces. Its size is given by its "number" (number 4 rebar has a diameter of ⅛ of an inch, number 5 a diameter of ⅝ inch, and so on).

stretcher A concrete block designed to be stacked between other blocks (not at an end or corner).

stucco A mix including portland cement, water, and other materials used as a troweled-on siding.

superplasticizer A material added to grout to thin it so it flows readily.

surface-bonding cement See surface bonding material.

surface-bonding material A material that is placed (usually troweled) onto the faces of a block wall to bind the blocks and (sometimes) provide an air and water barrier.

tension rod A steel rod running vertically through the center of a wall, tightened at the ends to provide reinforcing for the wall.

thermal mass effect The evening out of temperature swings inside a building as a result of having materials with large amounts of thermal mass.

thermal mass The ability of a material or wall to absorb and hold large quantities of heat or cold measured by the number of Btus of heat it absorbs (or releases) when its temperature rises (or falls) by 1 degree Fahrenheit.

thermal resistance The ability of a material or wall to slow the movement of heat through itself.

tie wire A piece of wire coiled around concrete forms or steel reinforcing bars to hold them a set distance apart during a concrete pour.

tie A piece of metal, wire, or plastic that connects pairs of concrete forms together to hold them in place against the force of the concrete inside during a pour.

tilt-up Using concrete walls that are shaped horizontally in forms or molds at the job site and then lifted into their final position.

traditional stucco See portland cement stucco.

U-block A special concrete block with cut-down webs, used mainly for courses that are to be filled with rebar and grout, usually to form bond beam or lintel.

Uniform Building Code The model code most widely used by states and localities in the West.

value segment The segment of buyers who want a home with the maximum floor space they can afford without using products or building practices that are below standard quality.

vapor permeance See water vapor permeance.

water vapor permeance The rate at which water vapor passes through a material, usually measured in perms.

web A cross piece in a block that connects the front and back face shells.

wet cup method A method of determining the water vapor permeance of a material by placing it over a container of water and measuring how fast water vapor passes through the material to the outside.

wet-mix shotcrete See wet shotcrete.

wet shotcrete Shotcrete mixed before being forced down the hose by the pump.

wind speed rating The highest speed of wind that a particular wall or structure is estimated to be capable of withstanding.

yard Slang for cubic yard, a measure of volume commonly used for concrete.

Zone See seismic zone.

Index

About the authors

Dr. Pieter VanderWerf is an Assistant Professor at the Boston University School of Management, where he teaches courses in business and conducts research into innovations in the construction industry. He has worked as a residential carpenter and remodeler, and has worked for three years at the NAHB Research Center, where he performed business evaluations of new building products and projects. He has also worked at various times for the Information Systems divisions of Haskins and Sells and Millipore Corporation, and he has been a consultant to the Portland Cement Association, the National Concrete Masonry Association, and several private corporations.

W. Keith Munsell has 22 years of experience in general contracting and real estate sales, management, and development. He is president and founder of Resource Concepts, Inc., a construction and development firm specializing in new single-family homes and townhouses and renovations and conversions of existing properties to residential use. He is a former officer in the Army Corps of Engineers, a realtor, a licensed broker, a certified property manager, a licensed construction supervisor, and a home improvement contractor. He is also an adjunct faculty member at Boston University and Boston College, where he teaches courses in real estate finance, management, and development. Mr. Munsell was awarded Boston University's Beckwith Prize for teaching excellence.